Into the Jet Age 1945–1975

Into the
Jet Age

Conflict and Change in Naval Aviation
1945–1975

An Oral History

Edited by Capt. E. T. Wooldridge, U.S. Navy (Ret.)
Foreword by John S. McCain

Naval Institute Press
Annapolis, Maryland

Library of Congress Cataloging-in-Publication Data

Into the jet age : conflict and change in naval aviation, 1945–1975 : an oral history / edited by E.T. Wooldridge : foreword by John S. McCain.
    p.   cm.
    Includes bibliographical references (p.   ) and index.
    ISBN 1-55750-932-8 (alk. paper)
    1. United States. Navy—Aviation—History. 2. Aeronautics, Military—United States—History. I. Wooldridge, E. T.
VG93.I54   1995
359.9'4'0973—dc20   94-41892

Printed in the United States of America on acid-free paper ∞
9 8 7 6 5 4 3 2
First printing

Unless otherwise indicated, all photographs are official U.S. Navy photographs and were obtained from the Library and Photographic Services Division of the U.S. Naval Institute.

Publisher's note: The descriptions of otherwise undocumented personal incidents and recollections of episodes and persons are as they appear in the memories of the individuals so credited. Every effort has been made to ensure correctness; inaccuracy, if it occurs, is regretted.

To those who fought to preserve our naval aviation heritage

Contents

CONTENTS

Foreword

The period from 1945 to 1975 was one of the most exciting and important times in the history of naval aviation. There has been no other period in history during which the changes in world order, technology, strategy, tactics, indeed almost every aspect of both civilian and military life, have been as rapid or as radical. The transformation of naval aviation, the evolution of aircraft carriers, the aircraft that flew from their decks, and the men who flew them are the subjects of this intriguing work.

World War II provided the first inkling that the ability to project power with aircraft would be a decisive factor in modern warfare. With the proper complement of men, machines, and weapons, the aircraft carrier became a key element in the protection of vital U.S. national interests worldwide.

Aircraft carriers developed into the most feared of any capital ships. From relatively small vessels with straight, wooden decks, aircraft carriers were modernized during this period into huge, powerful ships, capable of transiting any ocean of the world at speeds in excess of 30 knots, carrying a complement of up to ninety aircraft that could be launched and recovered simultaneously, in any weather, at any time of the day or night. Since the end of World War II, aircraft carriers and their crews have been called upon to project power scores of times. The men have always been ready and willing. The carriers have always been there.

If the navy did not lead the way into the jet age, it was certainly quick to realize the crucial importance of jet-powered aircraft. The leaders of naval aviation realized during the Korean War that jet aircraft would provide tremendous flexibility in operations, would offer more offensive firepower, and, indeed, were the way of the future.

Naval aircraft underwent a tremendous evolution during this period, with the development of jets, helicopters, and weapons systems. Radars were incorporated in naval aircraft to extend the range at which enemy aircraft could be detected and, with the development and integration of missiles, destroyed. Helicopters were transformed from a novelty into versatile and valuable platforms that could sink ships, hunt submarines, deploy SEALS and U.S. Marines, and rescue downed aircrews. The avionics and instrumentation that were developed during this period enabled operations around the clock and in almost any meteorological conditions.

But the development of new hardware is only part of the story, because the machines are worthless without individuals of talent and honor to operate them. Naval aviators who served during this period were some of the greatest of American heroes. They went from operating at maximum speeds of 350 mph in propeller-driven aircraft, to speeds in excess of 1,500 mph in multiengine jets. Naval aviators were among the early pioneers to break the sound barrier, and among the first men to go into space and land on the moon. Navy pilots crawled into untested vehicles and bravely flew them to their limits, and beyond, in an attempt to improve the combat capability and flight safety of carrier aviation. They braved the rigors of some of the most savage aerial combat ever waged. They withstood harsh weather, long deployments, bad food, and even the unmentionable horrors of detention, all in the selfless devotion to the duty of serving their country.

Change offers the prospect of great reward; it also poses great risk. Managing change for the better requires leaders of great vision and integrity. The naval officers who contributed to this historical volume are some of the finest leaders and aviators America has produced. Some of these men were my role models as a young aviator. I served with others and admired their courage, flying abilities, and common sense. I envied their judgment and savvy, for unlike most of them, I exercised poor headwork in intercepting a SAM over North Vietnam. But even that wasn't all bad, since I was able to lose weight and serve with Bill Lawrence in the process.

These admirable men led the navy through a tumultuous period. They stayed the course through two major downsizings of the military, two major conflicts, and countless encounters with the Congress of the United States. Their vision and ceaseless energy provided the proper mix of carriers, aircraft, weapons, aircrews, and sailors that helped America end the Cold War and maintain its position as the dominant military power in the world.

I thoroughly enjoyed this fine book, as I know you will. Tim Wooldridge, a top-notch naval aviator in his own right, has done an outstanding job putting together this collection of oral histories in a manner that provides a true depiction of the events of this period, as well as showing a glimpse of the personal travails and triumphs of these distinguished naval officers.

—Senator John S. McCain

Acknowledgments

I extend my sincere thanks to Paul Stillwell, Director of Oral History, U.S. Naval Institute (USNI), Annapolis, Maryland, who fully supported this project and placed the resources of the USNI Oral History Collection at the disposal of the National Air and Space Museum. Mary Beth Straight of the Library and Photographic Services, USNI, provided her usual excellent service in the selection of photographs. Dr. John T. Mason, founding father of the USNI Oral History Collection, oversaw the work involved in about one hundred thirty volumes of oral history and was ably assisted in the interviewing process by Paul Stillwell.

Members of the National Air and Space Museum staff who provided advice, encouragement, and assistance during the project include Nadya Makovenyi, Assistant Director for Exhibits and Public Spaces, Annette Newman of the Computer Services Division, and Mark Avino and Carolyn Russo of the NASM photo lab, who provided timely and highly professional support. I am particularly indebted to Adm. Arleigh A. Burke for his willingness to contribute his cogent and illuminating thoughts on the strategic importance of naval power in the latter half of the twentieth century. I am most grateful to Capt. G. G. O'Rourke, one of the true disciples of all-weather flying in the 1950s, and to those other planners, strategists, and pilots who fought the good fight, both in Washington and in combat at sea, to preserve our naval aviation heritage during those very challenging years following World War II and who had the foresight to leave their reminiscences for our continuing enlightenment and guidance.

Introduction

The U.S. Navy emerged from World War II unmatched in history for its offensive power and mobility. At the core of this invincible fleet was naval aviation, of which the key offensive element was the aircraft carrier. During the war, U.S. naval aviation expanded at an unprecedented rate, from a modest, untested peacetime force of one escort carrier, seven fleet carriers, and 1,774 combat aircraft to a fleet of ninety-nine carriers of all types, and 29,125 combat aircraft. The fleet of seventeen *Essex*-class carriers, which soon increased to twenty-four after the war, would undergo extensive modernization in the 1950s; eleven of these ships bore the burden of the carrier war in Korea, while seven of these saw combat during the Vietnam War.

As the various elements of naval aviation fought widely diverse forms of naval warfare over the vast expanses of two oceans, requirements for new war-fighting capabilities precipitated an infinite variety of technological changes and innovations. Advances in such areas as electronics, communications, weapons, aircraft power plants, and shipboard damage control systems were accompanied by parallel developments in operational procedures, doctrine, and training. In the Pacific, refueling and replenishing at sea were staged and choreographed to such a fine degree that task forces could be replenished on a grand scale under practically any circumstances, affording the navy the mobility necessary to carry the battle to the Japanese homeland.

All of this was accomplished with a virtually inexhaustible supply of resources—people, money, and materials—under the umbrella of a national will to win the war at any cost. After war's end, however, this unprecedented mix of offensive power and skilled human resources began to erode in alarming fashion. The abrupt shift from war to peace, followed quickly by the onset of the Cold War, precipitated periods of change so dramatic, intense, and revolutionary that naval aviation soon found itself in a fight for survival.

Sudden and rapid demobilization of the armed forces resulted in near chaos, uncertainty, and lack of purpose and direction. Within the aviation navy, carriers and squadrons were decommissioned at an astounding rate. At the same time, advancing technologies, such as jet propulsion, guided missiles, and atomic weapons, began to effect fundamental changes in our capabilities to wage war. The navy's immediate and pressing problem was to begin the evolutionary process of matching new technologies with obsolete weapons and doctrine, all the while maintaining an overseas presence and redefining roles and missions in the newly created National Defense Establishment. The battlefield shifted rapidly from the waters of the Atlantic and the Pacific to the halls of the Pentagon and the committee rooms of Congress.

In the fleet, pilots fought their own personal battles for survival. Seaplane pilots flying from battleships and cruisers on logistics or gunfire spotting missions realized their days were numbered. The Curtiss SC-1, last of the long line of ship-based scouts, left service in 1949, as catapults on the fantail were replaced by helicopters. By 1961, the airship, another reminder of the golden age of aviation, had also seen its missions absorbed by land-based patrol planes and helicopters. Eventually the last operational flying boat squadron of Martin P5Ms was decommissioned in 1966, their mission more efficiently accomplished by long-range land-based airplanes such as the Lockheed P2V and its successor, the P3V.

Such changes were logical and inevitable, if not always welcome, and were for the most part accomplished in a somewhat orderly fashion. However, the brunt of the transition to the jet age was borne by the carrier pilots, using the outmoded procedures and obsolete equipment of the war years. It is difficult for carrier pilots of later generations to appreciate the formidable emotional and physical tasks that faced the pilots of the early 1950s, whether they were flying jets or props. They were often inadequately trained and, in jet squadrons, sometimes led by pilots who did not understand what jets could or could not do, and often, through ignorance or stubbornness, placed their pilots in situations from which there was no safe way out. It was difficult enough for the new pilot in his first fleet squadron, fresh from Pensacola and jet transitional training; he still had the physical reactions and invincibility of youth to get him through the rough times, if he was good enough or lucky enough. Pity the unfortunate recalled reservist, a multiengine or seaplane pilot during the war, a family man suddenly thrust into flying situations that were totally foreign to him and that demanded skills he did not possess. It made little difference whether he was flying Hellcat or Corsair, Banshee or Panther. The catastrophic accident rate of the postwar decade was the worst in the history of peacetime naval aviation.

In the case of the jets, the heart of the problem was a combination of factors that included carrier landing approach speeds far in excess of those of the piston-engined planes of the war, resulting in drastically reduced reaction times on the part of the pilot and the Landing Signal Officer (LSO), slow acceleration time of the early jet engines, heavier aircraft with structural design deficiencies, poor flight deck illumination, and poor approach control facilities on board ship. Night flying off a straight deck carrier in the early jets frequently added a whole new meaning to the phrase "stark terror." All of the ingredients for an accident were present: an inexperienced young (or old) pilot attempting to land aboard ship at night in bad weather, flying on instruments at a very low altitude over the water in a descending turn on final approach, without radar control, while looking for the ghostlike figure of the LSO in his lighted flight suit, waving lighted paddles at an airplane he could barely see in the dark. It was the extreme test for man and machine— and naval aviation.

But eventually the solutions did come. By the late 1950s and early 1960s, naval aviation was moving into its fourth generation of tactical jet aircraft and critical changes had been made. From the British came the angled deck, steam catapult, and mirror landing system—later to be replaced by the Fresnel lens. The *Essex*-class carriers of World War II were modernized to accommodate high-performance jets. The so-called "super" carriers like the *Forrestal* arrived, and the nuclear-powered carrier was on the horizon.

The Replacement Air Group concept had been instituted, whereby each pilot destined for a fleet squadron was given an intensive four-month flight- and ground-training syllabus in the type of aircraft he would fly in the fleet. Finally—after years of frustration, doubt, and perseverance—aircrews, aircraft, and ships were at the same level of development and performance. Naval aviation was capable of truly all-weather air operations in a reasonably safe and proficient manner, and the crisis—and it had indeed been a crisis of the first magnitude—had passed.

The challenges associated with taking jet aviation to sea were aggravated by continuing requirements to maintain two carriers on station in the Mediterranean as a NATO commitment and two or three carriers in the Western Pacific. These task forces were frequently used to respond to crisis after crisis on the littorals of practically every sea and ocean of the world. Between 1945 and 1975, carrier task forces responded to crisis situations in such widespread locations as the Taiwan Strait, the Middle East, Africa, and the Caribbean, in addition to reacting to such events as the capture of the USS *Pueblo* in 1968, the loss of an EC-121 Constellation reconnaissance aircraft in 1969, and the capture of the merchant ship SS *Mayaguez* in 1975.

Twice in three decades violent peace gave way to so-called "limited wars"—Korea and Vietnam. In Korea, the navy adapted from the island-hopping strategy of the Pacific war to one of flight operations over a large landmass while the carrier steamed in a fairly stationary sanctuary offshore. Korea became a proving ground for carrier-based jet combat operations. Together, generations of young carrier pilots and senior commanders learned the complexities of operating jets at sea under combat conditions and, through trial and error, created the doctrine that became the foundation for modern sea-air operations.

In Korea, naval and air force aviators also made the agonizing, frustrating transition to a new way of fighting war—restraint, punctuated by artificial rules of engagement guaranteed to afford the enemy every possible advantage, to the detriment of our own pilots. It was a philosophy carried to a deplorable extreme by civilian leadership in the Vietnam War. In Vietnam, aircrews not only had to contend with politically motivated constraints on their actions; a new and deadly element of aerial warfare was introduced, for which U.S. pilots were not prepared—the surface-to-air missile. As a result, new tactics evolved and electronic warfare emerged as the most important technological development of the era.

Despite the obvious superiority in weaponry, the lack of any national will to win led to withdrawal from Vietnam in 1973. For nearly ten years carrier aviation had borne the burden of the navy's air action over Vietnam, and by the mid-1970s, men and machines were tired and worn out, money had become scarce, and drugs and racial problems plagued the fleet. A "hollow force" emerged. Yet requirements for overseas presence and responses to international crises persisted, and the Cold War continued unabated.

It is a tribute to the foresight, wisdom, and great courage of its senior military leadership that naval aviation survived the seemingly endless series of problems and challenges that faced the navy during the years immediately following World War II. Despite the traumatic effects of demobilization, reorganization, unification, crises at home and overseas, wars hot and cold, technological and scientific advances that appeared on the scene with bewildering frequency, and the social ills that beset the country and the navy, naval aviation survived and emerged better equipped, more capable, and more professional. In fact, by the late 1960s and early 1970s, U.S. naval aircraft and weapons could, by any objective criteria, be ranked as among the very best in the world.

Of all the major technological advances in aircraft and weapons systems capabilities in the postwar era, none was as dramatic, revolutionary, and imaginative as the first (and only) U.S. operational V/STOL (Vertical/Short Takeoff and Landing) aircraft—the AV-8 Harrier. Developed by Hawker Siddeley Company in Great Britain, and eventually produced in the United States as the AV-8B by McDonnell Douglas, the AV-8A Harrier was placed into service in 1971 by the Marine Corps as its primary ground attack aircraft. The Harrier's ability to take off and land vertically is afforded by a vectored-thrust turbofan engine, which makes the airplane particularly suitable for operations close to the front lines, as well as from air-capable ships such as the LPH amphibious assault ship (LPH).

Despite the obvious advantages of such an aircraft for the Marine Corps mission, the Harrier's road to success was tortuous and bumpy. Many dedicated and farsighted Marines fought countless battles in the Pentagon and on Capitol Hill to convince the many skeptics and detractors of the potential of this unusual weapons system. Marines like Lt. Gen. T. H. Miller, who was a colonel when he began his crusade for the Harrier in 1968, fought lonely battles, with little or no support from the other services. The Harrier went on to prove itself in combat, and in the 1990s, the radar-equipped Harrier II Plus model entered service, with night attack capabilities and vastly improved performance and systems reliability.

A testimonial to the high standards set for naval aviators, and navy test pilots in particular, was the key role played by naval aviators during the early days of the manned space program. Beginning with the selection of four U.S. Naval Test Pilot School graduates in April 1958—John Glenn, Wally Schirra, Alan Shepard, and Scott Carpenter—naval aviators made major contributions in the Mercury, Gemini, and Apollo programs, including the first American suborbital and orbital flights.

The period under discussion was one of transition for the world, for the nation, for the U.S. Navy, and for naval aviation. Indeed, there is an entire generation of naval officers, "brownshoes" and "blackshoes," who spent a career in "transition." Every community in naval aviation—antisubmarine warfare, transport, helicopter, early warning, patrol, and so on—went through its transition, some with relative ease, others in traumatic fashion. U.S. naval aviation covers an expanse of roles and missions that is impossible to define or address in any collection of reminiscences such as presented here. No single group of officers of any reasonable size could cover the gamut of land, sea, and air operations in which naval aviation has played a role.

Consequently, I have chosen to narrow the focus of this book primarily to the story of the aircraft carrier, although the discussions contained herein, particularly those concerned with policy and management decisions by the various chiefs of naval operations, their deputies, and bureau chiefs, had profound and lasting effects not only on carrier aviation, but on the entire U.S. Navy. Similarly, implied in discussions of international crises of the period—the Cuban missile crisis in 1962 for example—is the vital role played by every component of naval aviation, and the navy as a whole. In fact, maritime patrol squadrons involved in enforcing the quarantine flew more flights that came in daily contact with the adversary than did carrier aircrews flying mostly routine training flight operations from their stations south of Cuba.

The carrier becomes the principal focus of discussion for a variety of reasons. It played a predominant role as an instrument of diplomacy in peacetime and was the leading offensive weapon of the U.S. Navy in any conventional conflicts during the thirty-year period under consideration. Carrier aviation was the lightning rod, the focal point of interservice squabbles over roles and missions in its fight for survival as part of the navy's air arm. The question could be asked, "If other branches of naval aviation had been severely curtailed or eliminated, would the impact on the navy's capabilities to wage war or keep the peace during the period in question have been as critical as it would have been if carrier aviation had shriveled up and died?"

Perhaps never before have the reminiscences of such a distinguished group of naval officers been assembled under one cover. Some of these great men presented the case for naval aviation with conviction and determination to the "whiz kids" in the Pentagon and the skeptics in Congress; others supervised the design of the aircraft and the ships and shepherded them through the appropriate committees of Congress, into production, and on to the fleet. Others flew the aircraft—tested their flying qualities and performance, performed aerial demonstrations for the public, and in the ultimate test, engaged the enemy in aerial combat. A chosen few reached the pinnacle of their profession, and bore the burden of providing the leadership and wise council to carry the navy through many years of rough weather.

The reader may find two striking similarities in the problems that faced naval leaders during the postwar era and those that have emerged in the Department of Defense during the early and mid-1990s. First, many of the narratives in this book are extremely critical of the civilian leadership of the day—for failure to appreciate the unique capabilities of the navy, for duplicity, for overcentralization, and for failure to heed the advice and council of those senior people in uniform with years of experience in the fleet. Second, the leaders of the past had to contend with a number of divisive social issues—drugs, racial dissension, "liberal" lifestyles—some of which were merely controversial in nature while others, at the other extreme, threatened the good order and discipline of the navy.

As history has shown, our naval leaders of the past met these challenges and a better navy emerged. As the U.S. Navy moved into the decade of the 1990s, it again found itself beset on all sides by social issues, signs of deterioration of its moral fiber, uncertain leadership, and still yet another fight for the survival of naval aviation in the New World Order. A new generation of senior naval leadership was once again faced with the challenge of demonstrating the same degree of wisdom, intestinal fortitude, and good moral judgment so often reflected in these pages.

—E.T. Wooldridge

Into the Jet Age 1945–1975

IMAGES OF FLIGHT

The massive demobilization that immediately followed the end of World War II brought chaos, confusion, and uncertainty to the U.S. Navy. Nonetheless, for many naval aviators it was an enjoyable time. There was a plentiful supply of shiny new airplanes for everyone, and restrictions on where to go, how to get there, and what to do once you were there were about as lax as they would ever get. In time, however, the climate changed, as the 1950s became a turning point for naval aviation, particularly carrier aviation. Initially, the problems associated with taking jet aircraft to sea were not thoroughly understood by senior naval aviation leadership, at sea or ashore. Finally, when those problems became obvious, after years of fiascoes, close calls, and an alarming accident rate, the solutions were also not readily apparent. As all-weather operations became a reality, flying became a lot less fun, particularly when pilot experience, training, and self-confidence lagged far behind an overly zealous ship or squadron commanding officer's impulse to "Launch 'em!" on a dark and stormy night.

Equally less understood, but in many respects just as challenging, were helicopter operations in the shipboard environment. As with the carrier navy, improved operating procedures, equipment, and training came slowly, and frequently in response to catastrophe or mishap. Perhaps no type of aircraft in the navy inventory was as adaptable as the helicopter to a wide variety of missions in the decades that followed. From search and rescue and astronaut recovery to underway replenishment and mine countermeasures, the versatile helo and its dedicated, largely unrecognized, flight crews continued to find new ways to get the job done—whatever and wherever it happened to be.

As the Cold War became a fact of life in the immediate postwar years, and overseas deployments of aircraft carriers in the Mediterranean and the Western Pacific became an irrevocable commitment, a new phrase crept into the everyday vocabulary of the peacetime navy: tempo of operations. From the most senior commanders down to the firemen in the engine room, deployments of six to nine months became routine, as did extensions of deployments by a few days, a few weeks, or a few months. Morale suffered, and problems associated with transition to jets, parts shortages, aircrew proficiency, and training all became aggravated.

The decade of the fifties saw a bewildering array of second- and third-generation jets enter the fleet. As lessons learned in the fleet were fed back to the shore establishment, maintenance and flight crew training and survival techniques and equipment developed in World War II were improved dramatically. The casual approach to aircraft checkout and operational training at the squadron level became a thing of the past. Gone was the "kick the tire, light the fire" approach for the new squadron pilot.

In March 1958, the concept of the Replacement Air Group (RAG) was initiated as part of a sweeping reorganization of carrier aviation, creating a more permanent air group assignment to each carrier and providing for a permanent Replacement Air Group to be established on each coast. The RAGs would be responsible for the indoctrination of key maintenance personnel, tactical training of flight crews in fleet aircraft, and conducting special programs required for the introduction of new models of combat aircraft into the fleet.

In early 1961, another gigantic step forward in terms of standardization of operating procedures occurred when the first Naval Air Training and Operating Procedures Standardization (NATOPS) Manual was promulgated, prescribing standard operating procedures and flight instructions for each type of aircraft in the inventory.

Another significant milestone on the road to a safer aviation navy was the tremendous advances in the field of flight simulation. The Link trainer of the 1940s evolved into the Operational Flight Trainers (OFT), where mistakes could be made without the dreadful consequences of real-world accidents. Still later, Weapons Systems Trainers came along, and, with the arrival of the first night carrier landing simulator in 1972, flight crews could finally rehearse every conceivable emergency and operational situation that they could encounter in the air.

Safe and efficient operations on and around the carrier at sea during these formative years gradually improved. As the battles were waged in Washington for the survival of the carrier as a vital element of our nation's forward line of defense, measurable success was achieved in terms of pilot training, aircraft performance and carrier suitability, and aircrew survival equipment and procedures. Commensurate changes were made on board ship, as damage control systems and procedures, navigation systems and homing beacons, search radars and identification systems (IFF), radio communications, and radar approach control facilities became the focus of improvement programs. Automatic carrier landings were conducted with the F-4 Phantom and the F-8 Crusader in June 1963, a harbinger of times when most carrier-based aircraft would be equipped with the capability for the pilot to make "hands-off" landings on the ship. Of all these changes, however, the arrival of the steam catapult, mirror landing system, and angled deck in the 1950s—all British innovations—produced almost instant, recognizable improvements in the shipboard accident rate.

As an example of the many incremental, but very important, systems to be introduced on board ship, the Pilot Landing Aid Television (PLAT) System was installed on the USS *Coral Sea* in 1961. The system provided a videotape of every landing, and eventually every takeoff, and was extremely useful in the analysis of pilot landing and takeoff performance and accidents associated with flight deck operations.

The path to a safer, more effective, and more professional aviation navy was a bloody one, but by the early to mid-sixties, U.S. Navy aircraft, their weapons, and weapons systems were second to none in performance and mission versatility, and were employed extensively by free-world air forces around the globe. As proof, the navy began an all-out assault on many of the world's aircraft performance records. The F-4 Phantom II led the way with eight national and international world and class records in 1961–62. Other types of aircraft participating in the effort ranged from the Mach 2.0 A3J Vigilante to helicopters and amphibians. In the wake of the onslaught of technological changes of the 1960s, as if on cue, in August 1962 a navy airship made its last official flight at Lakehurst, New Jersey, ending a forty-five-year saga that began with the DN-1, the navy's first airship; an SP-5B Marlin flying boat made its last operational flight on 6 November 1967; and in July 1971, after years of valiant service in every conceivable role, the A-1 Skyraider, the venerable Spad, retired from active service. And to add a note of finality to the era of the piston-engined aviation navy, in February 1972 the Aviation Machinist Mate school for reciprocating engines was closed. An era had come to an end.

With each succeeding generation of military aircraft came dramatic increases in engine, aircraft, and weapons systems performance. Costs escalated in alarming fashion. In the early 1960s, during the "McNamara years," the push for commonality of parts, systems, and aircraft within the services resulted in the controversial TFX affair. The crux of the controversy was not a lack of agreement on the part of the navy or air force as to what designs would best meet the requirements of each service. Rather the problem lay in the refusal of civilian management in the Department of Defense to concede that years of military operational experience and technical expertise were an essential part of the military requirements process. Military judgment gave way to intellectual arrogance, deceit, and conspiracy, which, on the one hand, wasted money and time, and in the longer term, led to the tragedy that was the Vietnam War.

The TFX ultimately led to the navy Grumman F-14 Tomcat of the 1970s, and with the 1983 introduction of the F/A-18A, which had its genesis in the study of an experimental fighter/attack aircraft (VFAX) in 1973, naval aviation had progressed from an era where quantum advances in aircraft performance from one generation to another were commonplace, to an era where the performance characteristics of a new aircraft varied little from its predecessor. However, advances in computer technology brought vast improvements in weapons control systems, guided munitions, and of no little importance, the maintainability of the aircraft and its systems. The art and science of flying and fighting from a sea-based platform changed to a degree virtually inconceivable by pilots who cut their teeth on the nightfighter Hellcats and Corsairs of the 1950s.

As three decades of modernization, restructuring, and social change drew to a close, an event took place that, for naval aviators, old and young alike, would have a more disturbing, profound effect than the demise of seaplanes, blimps, or props. On 22 February 1974 Lt. (jg) Barbara Ann Allen became the U.S. Navy's first designated female pilot.

World War II ended too early for many junior officers like Gerry Miller. Designated a naval aviator as the war drew to a close, Miller brought to naval aviation the foresight, motivation, and love of flying so vital to the successful transition of the U.S. Navy into the jet age. His formative years as a naval aviator in the late 1940s were a naval aviator's dream—new carriers, new airplanes, few regulations. Miller's imagination was captured by the sounds and smells of jet fighters at air shows. His curiosity and ingenuity were taxed, and he became an instrument of change, as he sought and met head-on the challenge of adapting and refining this new technology. He came under the spell of visionaries like Bill Martin and Eddy Ewen, who pioneered in the development of night flying from aircraft carriers, and he also became a leading proponent of all-weather flying in the modern navy.

Francis Foley arrived on the helicopter scene during its infancy, bringing with him the unique perspective of a well-rounded aviator with experience in practically every type of aircraft in the navy. Admiral Foley remembers with nostalgia and obvious pride his experiences and the leadership he provided as the fleet, from ship captains to fleet commanders, discovered the seemingly unlimited potential of Foley's "choppers." Carrier aviation had its share of teething problems in the postwar era, but as Admiral Foley relates, making a night landing in poor visibility on the fantail of a pitching, rolling destroyer was no less fun—or demanding—for the helo pilot than the carrier pilot on the dimly lighted straight deck of an *Essex*-class carrier.

Naval aviators in carrier-based nightfighter squadrons who survived the decade immediately following World War II—the years of transition from props to jets, from straight decks to angled, from fair-weather flying to foul—can be forgiven if they occasionally look back on their experiences with a strange mixture of incredulity, nostalgia, anger, and relief at having survived the whole ordeal. In 1952, then-Lt. G. G. O'Rourke reported for duty to that granddaddy of all-weather fighter squadrons on the East Coast, Composite Squadron 4 (VC-4) at Atlantic City, New Jersey. With little previous all-weather flying experience and even less inclination toward pursuing a career in that direction, Jerry O'Rourke became a nightfighter, led a team of F3D Skyknights in combat in the Korean War, and went on to a distinguished career as an advocate and promoter of all-weather flying in the carrier navy. A master storyteller, Captain O'Rourke paints a graphic, sometimes frightening picture of the primitive world of nightfighters in the 1950s, a world little appreciated or understood by the majority of naval aviation leaders at the time. With his unusual insight into the human dimension, he presents an eloquent, well-deserved tribute to the aircrews who flew the night skies with the composite squadrons of that era.

The transition from Hellcat fighter pilot in the Pacific to OS2U seaplane pilot on a cruiser in the Mediterranean would be an unlikely, and unwelcome, career move for most. However, Ray Hawkins, one of the navy's leading aces with fourteen victories, took this temporary diversion in stride and was on hand to participate in the final farewell for the venerable Kingfishers and Seahawks as they were gradually replaced by the first generation of helicopters. Hawkins survived two cruises on board the USS *Portsmouth* with the Sixth Fleet to eventually join the navy's prestigious Flight Demonstration Team, the Blue Angels, and enjoyed two separate tours of duty with the team. During the first, he made the transition from the piston-engined Bearcat to the Grumman F9F Panther, the first jet flown by the Blues. After a combat tour during the Korean War, he rejoined the team as commanding officer in 1953; he affords the reader a behind-the-scenes look at the daily life of the team and a view from the cockpit as he describes the intricacies of an aerobatic routine with the Blues. Ray Hawkins's ejection through the canopy of a sweptwing Cougar at supersonic speed was not part of the show, and was his only bailout in thirty-one years of flying!

Before World War II, Jim Russell had been instrumental in the design of the *Essex*-class carriers, which contributed so much to the victory in the Pacific. In 1952, he became heavily involved in the development and fleet introduction of the angled deck, steam catapult, and mirror landing system, which made those same *Essex*-class carriers the bedrock of naval aviation until they could be replaced by sufficient numbers of larger conventional and nuclear-powered ships. Admiral Russell explains in lay terms the technical aspects of these British innovations and goes on to discuss in his usual articulate, candid manner, the aircraft that he considers to be the successes and failures in naval aviation during his tours of duty in the Department of the Navy and as chief of the Bureau of Aeronautics, and the trials and tribulations of shepherding the navy's aviation budget through Congress.

Flying in the fifties was lively and exciting, but the business of determining requirements, selling them to the Department of Defense and Congress, and fighting the battle of the budget was every bit as challenging, and demanded the same blend of courage, determination, and intelligence required by pilots in the fleet. Vice Adm. Robert B. Pirie—affectionately known in the naval aviation community as "the Beard"—assumed the job as DCNO (Air) in 1958, and fought valiantly and persistently as he shepherded two aircraft carriers and an assortment of aircraft through the bureaucracy. Some of these aircraft were still in the fleet some thirty years later—the Grumman A-6 Intruder and E-2 Hawkeye, the McDonnell Douglas F-4 Phantom II, and the Lockheed P-3 Orion. With honesty and the courage of his convictions, Admiral Pirie explores another dimension of the flying navy—the world of procurement, training, research and development, propulsion, and controversial issues such as the fate of lighter-than-air and the ill-conceived TFX airplane.

Three years after Admiral Pirie retired from the DCNO (Air) position, Vice Adm. Thomas F. Connolly burst upon the scene, bringing his own particular dynamic, aggressive style to the problems at hand. The TFX fiasco was still on the front burner and Tom Connolly wasted no time in making his presence felt, taking on the contractors and the defense establishment in turn. Admiral Connolly pulls no punches in his discourse on the behind-the-scenes skullduggery and political maneuvering in Washington as he fought for the "right kind of fighter" for the navy—ultimately the Grumman F-14 Tomcat. When he retired from the navy in 1971, "every aviation program was in apple pie shape; . . . the planes were the best that any tactical force in the world had or could lay hands on."

Vice Adm. Kent Lee had a slightly different perspective of the situation from his vantage point as head of the Naval Air Systems Command. When he took over in 1973, two years after Admiral Connolly's departure from DCNO (Air), the F-14 program had more than its share of problems, and Admiral Lee's account offers a less than rosy view of the program. He proceeds to unfold still another tale of controversy, contractual bickering, congressional hearings, and intraservice disagreements as he and his staff pushed the development of the joint service Lightweight Fighter project, which the DCNO (Air) staff, responsible for meeting the requirements of the fleet, did not want. Despite big divisions between the two staffs, the McDonnell Douglas F/A-18A emerged from the hassle. Whether the airplane becomes the "world beater" that Admiral Lee and the navy foresee, or fails, as a vocal segment of the naval aviation community predicts, will only become evident as the navy moves into the twenty-first century.

1

Transition to the Jet Age

Vice Admiral Gerald E. Miller

Gerald Edward Miller was born in Sheridan, Wyoming, on 1 July 1919, son of George E. and Nina (Smith) Miller. He graduated from Stadium High School, Tacoma, Washington, and in September 1936 enlisted in the U.S. Navy. In 1938 he entered the U.S. Naval Academy and was graduated and commissioned ensign on 19 December 1941. Designated a naval aviator in April 1945, he subsequently served in extensive aviation duties ashore and afloat, including squadron and staff combat tours during the Korean and Vietnam wars, as he advanced in rank to vice admiral, to date from 14 September 1970. Vice Admiral Miller became deputy director of the Joint Strategic Target Planning Staff for the Joint Chiefs of Staff on 15 July 1973 and served as such until his retirement, effective 1 September 1974.

From 24 August 1950 to 22 March 1951, Admiral Miller served as flag secretary to commander, Task Force 77, during operations against the enemy in the Korean theater, and in May 1952 he assumed command of Fighter Squadron 153. He led his squadron in combat operations in Korea, flying from the carrier Princeton *(CVA-37). Subsequent command tours included commander, Carrier Air Group 17, commanding officer, USS* Wrangell *(AE-12) and USS* Franklin D. Roosevelt *(CVA-42), commander,*

Vice Adm. Gerald E. Miller, commander, U.S. Sixth Fleet, 1971–73. Former carrier pilot, 1945–59.

Carrier Division 3, commander, U.S. Second Fleet, and commander, U.S. Sixth Fleet in the Mediterranean.

In addition to the Distinguished Service Medal with two Gold Stars, the Legion of Merit with three Gold Stars, the Distinguished Flying Cross, the Bronze Star with Combat "V," the Air Medal with Silver Star and two Gold Stars, the Navy Unit Commendation, and the Presidential Unit Citation Ribbon, Admiral Miller has numerous other campaign and service ribbons and several foreign decorations.

Soon after the end of World War II, I was fortunate to be assigned to a very fine air group, one of the oldest—Air Group 4, later converted to Air Group 1. It had the old Red Ripper squadron and the Top Hatters that had been around for years. We were based in Groton, Connecticut, near the submarine base, operating as a subsidiary of Quonset Point. New ships were coming out, and we were assigned to a brand-new aircraft carrier, the *Tarawa* (CV-40). The aircraft carrier had proved itself and it was now more accepted than ever. If you were a carrier aviator, you were really something.

It was unbelievable what we had in the way of assets. Here we had a brand-new aircraft carrier and a full air group running almost one hundred brand new airplanes. I rarely flew in an airplane that wasn't new for quite some time. Whenever we went anyplace, we had a full bag of good airplanes, and of course, we had competent people, well trained, even as we cut down the size of the navy.

Once the demobilization got under way, there was no stopping it. I remember that the navy went down to about two hundred fifty thousand men by 1949, from three million during the peak of the war. So the postwar years were spent trying to hold together, to get something constant established, while retrenching and cutting back. In the latter phases, most enlisted men belonged to the ship and repaired the airplanes. In late 1946, early 1947, we went back to the pre-World War II system, where all the men were assigned to the squadrons. You got control of your men and it gave everybody some experience in dealing with men. Now, all of a sudden, we had a maintenance division and personnel problems to solve. It was a restructuring, a real transition, trying to hold things together while we decided the size and nature of the navy.

To young people of my time, there was a lot going on. The Cleveland Air Races were an excellent collection point for a person to see the flying performances. They motivated the young to get into aviation, and those who were already in to do better. I still remember the 1948 races with two old training aircraft, manned by air force pilots, going across the field in Cleveland. They came out from below the hangar top level, doing a series of snap rolls, in formation, side by side. I thought that was one of the most spectacular things in the world. The stunt teams motivated you to fly better.

We flew everything; I eventually became qualified in all types of airplanes on our carrier. Later, in the middle of 1947 when we did start to have pilot shortages, I never knew sometimes what airplane I was going to fly when we were taking the air group aboard the carrier. It depended on who was short of pilots. We got to the point in 1947 where sometimes we had more airplanes than we had pilots.

At this stage we were starting to get more and more interested in all-weather flying and in night work. We recognized that we had to do a lot more of that, so we were conditioning ourselves for more instrument flying. One of the lessons of the war came from Rear Adm. Eddy Ewen and the *Independence* (CVL-22) operation at night. We started getting great lectures from guys like Bill Martin out of the Naval Air Test Center, Patuxent River, Maryland, in 1947 and 1948. Patuxent was really, to my way of thinking, a fountain of new ideas and concepts. The test pilot program was there and the best people were there—people like Bill Martin who had flown at night from the carriers during the war—and they were pushing all-weather flying. We were starting to get the F8F Bearcat fighter, which wasn't very good for instrument flying, but we had the F7F twin-engined Tigercat, which had more instrument capability than the others, so it became the nightfighter.

We were also starting into jet aviation in 1945. We had a squadron (VF-66) of Ryan FR-1 Fireballs, which was a combination reciprocal engine in the front end and a jet engine in the back. In 1948 we introduced our first carrier jet squadron, VF-5A, into San Diego with Pete Aurand as the commanding officer. That was a North American aircraft called the FJ-1 Fury. It had an engine that would only run for thirty hours, so they used to tow them out to the takeoff point and then start them up. When they came back from the short flight, they'd shut down the engine and get a tow back to the line, because they didn't want to waste the taxi time on the engine.

We were also starting to approach the limits of the straight deck aircraft carrier, and we were in a position where we had to do something about it. Of course, the British, very capably and timely, came out with the angled deck and the steam catapult in the early 1950s. We got those items just in time. We were starting to feel the pressures from jet engines, higher speeds, heavier aircraft, higher performance. The carrier of World War II had to change—and it did.

Eventually, my squadron deployed aboard the *Tarawa* from the East Coast through the Panama Canal to San Diego. Then we set out and we were in the Western Pacific for about a year, operating with the Marines in China, conducting simulated strike missions over North Korea. In the Pacific before the Korean War, we were reducing our presence, cutting down to a couple of aircraft carriers. By the time the Korean War started, we were down to one. I'd been all over North Korea before the Korean thing ever started. There was a lot of freedom to fly around different places—the Kwajalein area and all of those wartime flying spots. For the people who had not been in it, it was sort of reliving the war again. It was a great training period, kind of reflected glory.

The *Essex*-class ships like the *Tarawa* were fine for the aircraft that we had at the time. They were plenty big enough for us to land aboard. We used to kid about which runway to land on, the right or the left. We had many arresting wires, and the performance of the aircraft was sufficient in those days so that it was a piece of cake. There was no particular problem.

The *Midway* (CVA-41) and *Roosevelt* (CVA-42) came out in 1945, and the *Coral Sea* (CVA-43) a little later. If you got to one of the air groups on those ships, you were really leading the pack—steel decks instead of wooden decks. They were the ultimate. The main difference in combat capabilities between the two ships was that one carried more airplanes than the other, and it would have sustained more damage. Also, the accident rate could be reduced, because of the larger size. We ran into trouble with the *Essex* when we got into jet aircraft and still had the straight deck. I did that in Korea, and it was not easy; it was a marginal program.

As an example, the first year in the Korean campaign, in 1950–51, we had the F9F-2 Panther aircraft, which was the first jet that we really deployed in any numbers. I was on the task force commander's staff for that first year. We had been out there quite a while, and were going to change air groups from one aircraft carrier to the other. We were going to send one ship back to the States, but we were going to swap the air groups from one ship to another. So we very carefully picked the best airplanes of the two air groups. We had four jet squadrons, with sixteen birds in each one—about sixty-four jet fighters to make the swap. We tried to pull this switch in the air off Yokosuka, Japan.

The weather was bad over the beach. We didn't have a good shore base to go into in case of trouble, and the staff—and I was a member of the staff—didn't do a very good job in setting this thing up. We ran into one of those unfortunate conditions when the wind was from one direction and the sea from another, with quite a swell, and the decks started to pitch. We didn't put all 130 of the planes in the air simultaneously. We put them up, fortunately, a few at a time, but we had a lot of these fighters in the air. Then we tried to bring them down and it was a tough job getting them on board. They were running out of fuel and there was no base on the beach to send them to. We had to get them back on board those two carriers, and we broke up those planes in some numbers.

It was awful. It was so bad, I can still remember the admiral walking over to the opposite side of the bridge, putting his head down on his hands and shaking. It was so bad he couldn't even get mad. It was a horrible mess. Well, that was all because of the size of the ship, the nature of the airplanes, and straight deck operations. We started from debacles of that kind to get something better. It was at this stage that we started talking about qualifying pilots for carrier landings by "running through the deck." Instead of having barricades in front of us, during the qualification phase, the deck would be kept clear up forward. You would come in with hook up, and do a touch-and-go landing. You landed on the carrier, but just kept rolling, taking off from the front end without slowing down.

In early 1953 I vigorously pushed the staff in San Diego to try to let us do that with my squadron, to try to cut down on the landing accidents and get people some experience with bringing jet airplanes down on the deck. I was not allowed to do it, but a couple of months later, other squadrons were given permission and a whole new procedure was in vogue. You were allowed to come down and land, but if you did not catch a wire, you went up the deck and took off again. If you did catch a wire, it was a normal landing. About this time, the British came along with the angled deck idea, which was just perfect. But we all went to Korea on straight decks. We worried about jet landings, with a whole group of airplanes parked in front of the barriers. If you didn't catch a wire or the barricade, you could go into that mess up forward, and that happened occasionally.

Considering the upheaval in the navy caused by demobilization and the introduction of new technologies, it's amazing that we kept together as much as we did. We decommissioned ships and reevaluated, trying to get some new programs going. We worried, but we did proceed with the jet program. We had the night program progressing ahead in aviation. We didn't do much in the surface navy at all; the emphasis was all on airplanes, and of course, this was the stage when we started getting interested in nuclear power in the submarine force. I think in balance that we were probably doing pretty well.

2

Doing It in Rotation

Rear Admiral Francis D. Foley

Francis Drake Foley was born on 4 July 1910 in Dorchester, Massachusetts. On 2 June 1932 he graduated from the U.S. Naval Academy and after designation as a naval aviator on 1 February 1936 he served in various squadrons ashore and at sea until the summer of 1942, when he joined the USS Hornet *(CV-8) as air operations officer. After the* Hornet *was lost in the Battle of the Santa Cruz Islands, he was ordered to duty as assistant operations officer on the staff of commander, Task Force 65. He later served at Guadalcanal on the staffs of commander, Air, Solomons, and commander, Fleet Air, South Pacific. Between 1943 and 1945 he was head of the Officer Flying Section, Office of the Chief of Naval Operations.*

After World War II, Admiral Foley served on board the carrier Franklin D. Roosevelt *(CVA-42) and on the staff of the chief of Naval Air Training, Pensacola, Florida; he assumed command of Helicopter Squadron 2 in July 1949. Following tours of staff duty ashore, he commanded the seaplane tender* Salisbury Sound *(AV-13), flagship of the commander, U.S. Taiwan Patrol Force, from August 1955 until October 1956, and took command of the attack carrier* Shangri La *(CVA-38) in November 1956. Subsequent tours of duty included staff of commander in chief, Pacific; commander, Carrier Division 1; assistant chief of the Bureau of Naval Weapons for*

Comdr. Francis D. Foley, commanding officer, Helicopter Squadron 2, 1949–50. (Courtesy Rear Adm. Francis D. Foley, USN [Ret.])

Program Management; deputy assistant chief of staff, Plans and Policy Division, Supreme Headquarters, Allied Powers, Europe; assistant chief of naval operations (Fleet Operations and Readiness); commandant of the Third Naval District, in New York City; and senior member of the United Nations Command Military Armistice Commission, Seoul, Korea. He was placed on the retired list on 1 July 1972.

Admiral Foley's awards include the Legion of Merit, the Bronze Star with Combat "V," the Navy Commendation Medal with Combat "V," the Joint Services Commendation Medal, and the Chou Cross of South Korea.

In July 1949 I assumed command of Helicopter Squadron 2 (HU-2) at Lakehurst, New Jersey, after two years on the staff of the chief of Naval Air Training in Pensacola. The navy then had only two such squadrons, HU-1 and HU-2. As the senior naval aviator afloat at Lakehurst, I had a good set of quarters and delightful relations with all the lighter-than-air people on the base.

My squadron was keeping Lakehurst alive. HU-2 was a very large squadron; we had twenty-five Sikorsky HO3Ss, twelve Bell HTLs, plus two "flying bananas," the Piasecki HRPs, plus about four hundred men, and eighty officers. We also had flight students from the navy, army, air force, marines, New Jersey State Police, and the Federal Aviation Administration (FAA). They all had to learn how to fly helos.

Of course, I had to learn how to fly them too, so they put me through a two-week transitional course in helicopters as soon as I got there. I flew morning and afternoon, a couple of flights a day. I received my designation as a heli-copter pilot, and I guess I was within the first one hundred in the navy. It was relatively easy for me as an experienced aviator to learn to fly it.

Most of the helicopter pilots in the navy were in the squadron right then. Some were in HU-1, which was smaller, out on the West Coast, but HU-1 didn't do training like we did. We not only did training for the pilots, but for enlisted men also, and we furnished all the helicopter support for the Atlantic Fleet, including the Sixth Fleet in the Mediterranean. At one time I had twenty-three different units scattered around on carriers, cruisers, battle-ships, icebreakers, and survey ships. We were up on the Distant Early Warning (DEW) Line with a helicopter on floats, over in the Persian Gulf on an oceanographic survey ship, and down off the coast of Venezuela, around Aruba, with another one. Usually the aircraft that we sent out to the fleet were the HO3Ss. We even kept a small unit of three choppers down in Norfolk in those days just to take care of emergency demands in the Norfolk area, because of so much fleet concentration of big ships.

Four of us flew out in our Beechcraft to Los Angeles, and spent four or five days at the Los Angeles airport with Los Angeles Airways, which was flying the same chopper on the airmail routes to the satellite cities within about fifty miles. They were sending airmail from their post offices into the L.A. airport, flying quite a bit on instruments out there because of the visi-bility conditions. It was all low-altitude, landing on the rooftops of many of the post offices. I flew the whole route with the Los Angeles Airways—every stop with them. We compared notes, and while we were doing the business for the fleet back at Lakehurst, they were doing the same thing, practically, out there in L.A. So it was a very, very fruitful exchange that we had out there with them.

We did the initial transitional training in the HTL, which was a little Bell, two-place chopper, with an instructor and a student side-by-side. Most of the training was done right around the station—landings and takeoffs, patterns, figures of eight, squares, and diamonds, and so forth, laid out on the field. You'd go out and get about four or five feet off the ground, just out of ground

During the early days of U.S. Navy helicopter operations at Lakehurst, N.J., 1948 (left to right): Piasecki HRP-1, Goodyear blimp, Sikorsky HO3S-1, Bell HTL-1.

effect, and just fly your aircraft around—back it around, go sideways, forward—learning how to take off from a very congested area, how to take off downwind and get back into the wind in a hurry, things like that. Then we had to learn how to rescue people. When we shifted into the HO3S, the instructor had to ride in the back seat and coach you with earphones. But we learned how to use the hoists and rescue people.

The helicopters operated at night during that period. We had night flying from the carriers, battleships, and cruisers all the time, and at Lakehurst we were having night flying about once a week. Fortunately, the HO3S was a remarkably reliable machine. It couldn't carry a great deal—a pilot and normally three passengers, and maybe one or two extra—that was about it. Our availability normally ran close to 90 percent. For instance, aboard a carrier, cruiser, or a battleship, we could only have one helicopter, and, particularly on the carrier, that thing had to be ready to go all the time for flight operations, and then they used it for anything else of importance. Availability was one of the highest in the navy, consistently.

The HTL, the little Bell, was also an extremely reliable little machine, and that's the one we used for the Coast Guard for their icebreakers. We would put them on floats and send them up to Boston, where they would embark in a cutter and go up to replenish the DEW Line. The choppers were remarkable for the icebreakers in permitting them to navigate more efficiently in the ice. They could go out and scout and see what the formations of the ice were, and clue the conning officer on the cutter as to what course to take. So they were able to cover a lot more mileage in a day when they were under way with a chopper to guide them.

We found the same thing with the hydrographic survey. The choppers were able to go around and hover over the pilings or markers where they had instrumentation, taking readings while hovering right there alongside. They could read the instruments from five feet away, just get right on top of the thing and do it. They speeded up the business of the hydrographic survey, and, of course, they could do soundings and things like that if they wanted to. We had the New Jersey State Police enthusiastic about the traffic control possibilities of the chopper, and also the rangers for spotting forest fires.

Frequently the HO3S was landed on a forward turret of a battleship or cruiser, although we wanted to get it onto the fantail which would be more convenient. Up on top of a turret was a pretty hair-raising place to be. It started out on the turret because it was the only place that was completely clear. It was just so chopped up with 40-millimeter guns and stuff like that around the decks, they didn't have the space. It helped to get the catapults off the fantail also.

The air traffic people who controlled all the other air traffic at sea controlled the choppers, too, because for every flight operation from the carrier, they were the plane guards. They were always the first thing off and the last thing back on deck, so they were in close contact with flight control in the carrier. We also would have people in a battleship or a cruiser who would control the flights there. We would set up a control center, and generally speaking, they'd try to find a spot for it where they could see the helo instead of having to watch it on a radar. We would get somebody who was a pilot as controller. When we sent our units off, they always had two pilots and five enlisted men, so one pilot was almost always on the ship, and the other pilot was flying the chopper.

Once one of my choppers was landing on a light carrier, a CVL, and one of the flight deck crewmen put a hold-down on the chopper. He was going to put three of them on. The pilot decided that the ship was pitching and rolling so much that he would have to come back. He didn't know that a hold-down was already on, and he tried to take off, but the hold-down held him and, boy, that resulted in a lot of injuries; half a dozen people were hurt when a damaged rotor blade flew all over the place. Our signals between the pilot and the flight deck people were not adequate, so we had to go to general quarters about that one. That was the only accident we had in the squadron while I was there, thanks to a number of safety precautions written before I got the squadron. But you live and learn, and you have accidents, and you have to rewrite the book every once in a while, and this was a new one.

The ships were very cooperative, probably because they found the helicopters so useful. They were all very anxious to have them. The ships gave us inputs also on how the pilots were doing, and of course, when the people came back, they always sent a fitness report on them. I never remember anyone getting an unsatisfactory fitness report—no helicopter pilot. This is an interesting contrast to twenty-five years earlier when aviation was so unwelcome on the battleships and cruisers.

While I had that command, we were beginning to feel the effects of the Korean War buildup, particularly when I sent the choppers to Korea. They were among the first to be in Korea. I had to send six helicopters; they put them in flying boxcars, flew them out to the West Coast and then took them in the *Philippine Sea* (CVA-47) to Korea. Then it was obvious that if we were going to have this big helicopter expansion, they were going to have to expand not only the production base, but also the pilot and mechanic training. So they decided to put the helicopter pilot training, with our assistance, in Pensacola. We picked out Ellyson Field as the helicopter training field, close to the main station at Pensacola. For the technical training, my people went out to Memphis and established a training course with the syllabus that we had developed.

Then we started farming people out. I had to let a lot of my instructors go to Pensacola or to Memphis. The reason for putting in a training facility at Lakehurst in the first place was just to keep that station going, but it had a lot of advantages for experimental research and development too, so that was good.

In 1950, after I made captain, I was informed that I would have to give up my squadron command and go to Washington to the Office of the Chief of Naval Operations as head of the Aviation Personnel Programs Section. Since I was also a helicopter "expert," I was supposed to help in planning the helicopter business; I was going to have two jobs.

Of course, because of the Korean War the helicopter had suddenly become big business. We needed a lot of bodies in the helicopter field, and we preferred that they be jet-qualified, preferably prop-qualified, naval aviators before they transitioned into helicopters. We did not give people their wings as helicopter pilots. We thought that the people should have the experience in the regular fixed-wing heavy aircraft. All of us that were in it in those days had that background, and we felt comfortable with it. Part of our thinking was that a person wouldn't necessarily stay in helicopters. Maybe the helicopters would not prove interesting enough to keep people in; we actually didn't know how far they were going to go. However, we could visualize all kinds of useful things that a helicopter could do, as they were doing in my squadron even in those days, and we were right.

The navy was beginning to get into specialization at that time, where you could figure a guy would spend a whole career in one type. Of course, in my day, the active flying days, the diversification was very, very important. You really wanted to get your tickets in everything that you could—fighters, dive bombers, torpedo planes, and patrol planes, both land and sea. Of course, everybody still has that idea that he is basically a fighter pilot at heart. I know that during my career I've flown fifty-six different types of airplanes and choppers; eight of them were fighters.

While I was in this aviation plans business, one thing that we worked on very, very hard was to try to get the flight load at Pensacola straightened out for the long term. We did a lot of calculations and came up with a figure that would smooth out the peaks and valleys in the pilot input into the fleet and into the Naval Air Reserve Training business. Of course, things were complicated a little bit by the fact that the navy was getting out of the air transport and seaplane business.

I must admit that we did a lot of conniving about people. We were under very close scrutiny for money and having to justify every last officer and man and it was very difficult. We tried to bury billets in various unsuspected spots so that we would have people when we needed them desperately elsewhere. Just by placing a few extra billets here and there, you could build up a bank to make allowance for something crucial.

Although there was a real stringency on funds, that faucet opened up as the war came on. I think that the Korean War that was going on then gave us a lot more leverage than we would have had otherwise. The need to man aircraft carriers grew as they came out of mothballs; we had to get people, almost create people.

I think the Naval Air Reserve was very successful at keeping people up to snuff as far as being able to come back into the fleet. Usually the people who went into the Naval Air Reserve were people who loved to fly; they wanted to keep up their affiliation with the navy. They were loyal to the navy. They were anxious to get out and prove that they could do as well as the fleet pilots, and of course, in many cases they could do better, because they had more experience. A lot of them were either private pilots or airline people, as they are today, although the Naval Air Reserve, I believe, is much smaller now than it used to be. At one time, I think during my time in OpNav, it seems to me that we had six thousand people in the Naval Air Reserve who were actively plying their trade. It seems to me that there were something like twenty squadrons taken out of the Naval Air Reserve and sent to Korea.

I think that the most important contribution I made during that time in the Department of the Navy was to convince higher authority, all the way up to the secretary of defense, that we needed a stabilized pipeline in our Naval Air Training Command so we wouldn't have peaks and valleys in the pilot output and the amount of background they had. It seems to me that the pilot training program had been down as low as six hundred a year, and that wasn't nearly enough. It was proven to us immediately by the Korean situation. We just didn't have the people in the age or rank brackets that we needed. So we worked very, very hard on this thing and finally came out with a number that, as I remember it, was around 1,125 or 1,150 pilots per year. The idea was that the excess, after they had become fleet qualified and so forth, would go into the Naval Air Reserve Training Command, into squadrons, if possible, and keep that Naval Air Reserve alive for a fallback position, because it had proven to be so invaluable in the Korean War. We fought and fought and fought over those numbers with all the people up and down the line and everybody thought it was maybe a little too rich. We did attrition figures and projections on what it was going to take to sustain the fleet and the Naval Air Reserve, and so about 1,150 is the number we came up with, and they bought it.

When they bought it, I thought, "Well, boy, I have accomplished something."

3

We Get Ours at Night

Captain Gerald G. O'Rourke

Gerald Gerrard O'Rourke was born on 13 September 1924 in Brooklyn, New York, son of John J. and Geraldine Gerrard (Reilly) O'Rourke, an ex-yeomanette of World War I, and grandson of John A. O'Rourke, ex-chief boilermaker of the Brooklyn Navy Shipyard in the 1890s. He graduated from the U.S. Naval Academy in 1944 at age nineteen, and served on board the USS Hancock *(CV-19) as a Boiler Division ensign during the last year of World War II in the Pacific. He was designated a naval aviator in 1947 and flew F8F Bearcat fighters in Fighter Squadron 8A from several Atlantic Fleet carriers; he served as the legal officer of NAAS Whiting Field, Milton, Florida.*

In 1953, during the final stages of the Korean conflict, he served as officer-in-charge of a Composite Squadron 4 (VC-4) nightfighter detachment on board the USS Lake Champlain *(CVA-39) and ashore with Marine Night Fighter Squadron 513 (VMF(N)-513). Later tours of duty included the Naval Air Test Center, Patuxent River, Maryland, aide to commander, U.S. Sixth Fleet, officer-in-charge of Fighter Squadron 101 Detachment Alfa during the introduction of the F-4 Phantom II, commanding officer, Fighter Squadron 102 during the Cuban missile crisis of 1962, OP-05W in the Pentagon, commanding officer, USS* Wrangell *(AE-12), operations officer,*

WE GET OURS AT NIGHT

Comdr. G. G. O'Rourke, c. 1961. Former officer-in-charge of Composite Squadron 4 Det. 44N in Korea, 1953. (Courtesy Capt. G. G. O'Rourke, USN [Ret.])

staff of commander, U.S. Sixth Fleet, and commanding officer, USS Independence *(CV-62). Over the final three years of his career, he returned to OP-05W and led the Navy Fighter Study in their important efforts on behalf of the F-14 Tomcat, the Lightweight Fighter (later the F/A-18), and the Navy's embryo Vertical/Short Takeoff and Landing (V/STOL) program. Captain O'Rourke is a graduate of the U.S. Navy Test Pilot School and the National War College and holds an M.A. (Education) from Stanford University. He has written extensively on aviation and the aircraft carrier, primarily for the U.S. Naval Institute, where he once was a member of their Board of Control and was selected as their Author of Merit for 1976.*

Captain O'Rourke's decorations include the Legion of Merit with Gold Star, the Bronze Star, the Air Medal, the Navy Commendation Medal, and the Pacific Area Campaign Ribbon with five combat stars.

The early 1950s were confusing times for seagoing airmen. Although *Essex*-class aircraft carriers had contributed mightily to the victory in the Pacific, only a handful had escaped the massive mothballing of the late 1940s, amid interservice battles over roles and missions. A number of carriers were hastily reactivated without significant modernization for the Korean War and were staffed with recalled reservists who were equally behind the times. The first of the jets had arrived, and so too had the need for operating combat aircraft from ships around the clock in any kind of weather. Carrier aviation had become a turbulent mixture of the old and the new, led by jet fighters with fantastically better performance than their prop-powered predecessors. Flying them off the wooden flight decks of the *Essex* carriers was a challenging business even in the best of times—a calm sea and bright sunshine.

While the advent of the new jets was enthusiastically welcomed, new all-weather needs were only grudgingly accommodated. Naval aviation's planning for the simultaneous introduction of jets and all-weather flying was essentially half-baked. On the one hand, the prop-to-jet transition for the regular carrier-based squadrons, led by the fighters, was accomplished in a formal, well-organized, highly structured manner in a four-to-six-month period ashore, after which the squadron resumed its regular place within the air group assigned to each of the carriers.

This entire transitional evolution was very straightforward; schedules were rigorous; both pilots and ground crews received intensive specialized training; shortcomings of any sort drew immediate command attention, and stiff operational readiness tests were used as the "final exams." On the other hand, unfortunately, flying at night or in poor weather was given short shrift. There was, quite simply, not enough time, and besides, "That's not our mission—that's what the VC-3 and VC-4 guys do!" Who were these VC-3/VC-4 guys?

The task of Composite Squadron 4 (VC-4) was to do it in any weather, day or night. This mission was reflected in the squadron motto—*Nox Mea Auxiliatrix Est*. "Night is my ally" was close to the proper translation, but "We get ours at night" had a better ring to it and best suited the personality of the squadron. The genesis of VC-4 is a long story, going back to late in World War II, when a huge effort was begun to transition all carrier aviation to night and all-weather operations. This grand "master plan," at least for the Atlantic Fleet, involved successive, increasingly demanding pilot training periods at Key West, Florida, Atlantic City, New Jersey, and Quonset Point, Rhode Island, with the ultimate test a course at Argentia, Newfoundland. Postwar cutbacks had left only vestigial remains of this plan at Key West, where only very basic night and radar familiarization was possible, and at Atlantic City, where both operational training and actual deployment of small all-weather fighter and attack detachments ("dets") had to be provided by "composite squadrons"—VC-4 and VC-33 respectively.

The first of many VC-4 anachronisms was that it was, in reality, neither a "squadron" nor "composite." Its commanding officer was a fairly senior captain, not the lieutenant commander or commander normal to a regular squadron. Its normal allowance of personnel included about a thousand men and one hundred fifty officers, compared to a complement of two hundred men and about twenty officers in a "normal" squadron. VC-4, moreover, never itself went to sea, but at any given time, a half-dozen VC-4 four-plane, five- or six-pilot, forty-five-men dets could be found aboard various carriers all over the world!

The det, as a distinct organizational entity, was an anachronism in itself. Aboard ship, it was not a squadron, as were the other organizational entities around it, but merely a segment of a very large, but very distant, squadron (which itself was not really a "squadron"). The det's leader was an OinC, or officer-in-charge, and his rank was anything from a lieutenant commander down to a lieutenant (jg). The OinC's appointment to his exalted position was not the result of a formal board of selection, but the cursory action of VC-4's commanding officer, generally to satisfy only two mandatory requirements—availability and successful completion of eight night carrier qualification landings in the type of aircraft involved. His job carried very limited authority, little initial prestige, but lots of responsibility.

Aboard ship, his immediate "boss" was titularly the air group commander (CAG). In practice, some CAGs were great bosses, while others were not, at least initially. The OinC was usually assigned five or six seats in the rear of a squadron Ready Room for flight briefings, but he had to scrabble hard for a fair share of maintenance and berthing spaces, and make his two-man stateroom double as the det office. In general, most VC-4 OinCs were not on any career path other than survival, yet all too often they were called upon to make critical decisions on profound tactical and training issues.

The next VC-4 anachronism was that almost all of its Atlantic City efforts were involved with training of pilots in all-weather flying, fighter tactics, radar operations, and launching from and landing aboard a carrier at any time in any weather in at least one of the half-dozen or so different aircraft types (prop and jet) the squadron maintained and deployed. Added to this most basic mission were the comparable training for aircrew radar operators (ROs) who flew in the F3D; training for Ground Control Intercept operators (both officer and enlisted, from carriers in port); transitional training for pilots, aircrews, ground crews, and controllers moving from props to jets; and training of embryo Landing Signal Officers (LSOs, or "Paddles"). Then, too, because one or another of the dets at sea frequently suffered a mishap with an aircraft, a pilot and/or an appropriately configured replacement aircraft had to be immediately available for dispatch to some distant point.

Probably the oddest anachronism was that VC-4, for reasons that escape rationality, had also been saddled with providing some dets whose specialty was launching atom bombs onto distant targets and escaping their own demise in the process by performing a "loft" maneuver (essentially a tricky, ground-hugging Immelmann, in which the pilot performed a half loop, rolling out at the top to escape in the opposite direction). These so-called "Bravo" dets (they flew specially modified F2H-2Bs) were supposedly wrapped in airtight security, but could never escape the wonderment of the rest of VC-4 pilots, all of whom kept asking, "Why is this kookie idea a VC-4 mission?"

The ultimate VC-4 anachronism was its critical jet transitional training "program," which was distinctly an informal affair, notably unlike the extensive, thorough, formal education given elsewhere to the regular carrier "day fighter" squadrons. VC-4's "instructor pilots" were most often just other guys who had survived their own first jet rides a week or two before!

For second-tour pilots like me who had the experience of several previous carrier deployments behind them, the transitional stuff was exciting and viewed as a wonderful personal challenge. But for the kids newly arrived at VC-4 from the all-prop training command and a short night/radar course in props at Key West, sporting only about four hundred hours of total flight time, the simultaneous introduction to night, weather, and jets, all done in an informal, casual manner, presented a real exercise in personal survival. This demanding environment, as might well have been expected, exacted a very heavy toll in fatal accidents.

The most obvious oddity in odd VC-4 was its plethora of aircraft types, variants of those types, and few-of-a-kind subvariants of the variants. Props included ancient but radar-equipped F6F-5N Hellcats, plain-Jane radarless F4U Corsairs, F4U-5Ns, and a handful of F4U-5NLs, all with a pretty fair but rapidly aging radar, and a couple of SNB twin-Beechcrafts used for logistics and dubbed "Short Haul Interstate Transport Airline." In jets, there were a few plain-Jane radarless F2H-2 Banshees, a few F2H-2Ns that had the F4U's radar in the nose, F2H-3s and F2H-4s—"big Banshees" with a very lousy and a very outstanding radar respectively, the F3D-2 twin-place ("built from the keel up") nightfighter with very slow speeds but three good but fantastically hard-to-maintain radars, and a few two-place Lockheed TV-2 trainers.

Throughout its most productive years, VC-4's total stable numbered roughly seventy to eighty airplanes, with about one-third of them off on deployments at any given point in time. During this period, this same composite squadron scheme was also used at other air stations on both coasts to provide dets specialized in night light attack, heavy attack, airborne early warning, and photo reconnaissance, and some years later, for both high-tech electronic warfare and low-tech utility helicopters as well.

In retrospect, the scheme worked quite well in those situations where highly specialized warfare needs unexpectedly arose and could not quickly be fulfilled by the normal, well-established training programs. The primary weakness of the scheme was that it provided only a temporary solution, primarily useful for bridging the gap until the carrier's regular squadrons could be trained and equipped to handle the new task. In the case of all-weather fighters, the composite squadron scheme was vastly overextended, well beyond a full decade (1946–58), and was vastly complicated by the simultaneous transition from props to jets. For VC-4 itself, the late 1940s were reasonably static, with unique, but reasonably predictable duties to be performed. With the arrival of the jets in the early 1950s, and with the growing navy-wide need for all-weather, around-the-clock, around-the-globe, war-fighting capabilities, VC-4's unique characteristics thrust it to the forefront of carrier aviation.

Meteorology, geography, and personalities also had a lot to do with VC-4's unique flavor. The prevalent Atlantic City flying weather, at least in winter and spring, is best described as "continuously lousy," with plenty of rain to fuel the limitless dense fogs rising off the South Jersey bogs. The recall of many naval aviation reserve aviators during the Korean War came with mixed blessings for all the VC squadrons, and in particular for VC-4. Most of these merged well into the system as full-fledged "second-tour" pilots, and performed magnificently after barely minimal training. Others, however, had the seniority and the willingness, but neither the experience nor the talents, to contribute significantly in what had become a very high-tech business. Most of this latter group readily accepted the situation, openly eschewed serving in any deployed det, and applied themselves to a basic training task or a role in flying utility and logistical missions. The remainder of this particular group, however, presented essentially insoluble organizational and morale problems. The accusation was often heard in happy hour conversations that VC-4 was being used by the bureaucrats as a "dumping ground" for both dubious aviators and questionable aircraft that were essentially unassignable elsewhere.

VC-4's accident rate, when compared to any regular squadron or air group, was, quite expectedly, very poor. Just about every VC-4 mishap or tragedy could be traced directly to environmental factors, or indirectly to a lack of understanding of environmental factors on the part of some supporting personnel. After a deployment or two, every nightfighter came to realize that most of the accidents were caused by errors of judgment on the part of personnel on the ground or aboard a ship. Accordingly, he developed a very suspicious nature, and couldn't help feeling that everyone else in the navy, and Mother Nature as well, was conspiring to kill *him!*

So VC-4 was a weird outfit. By all standards, it should have been a notorious failure. Yet this weird outfit, along with its West Coast counterpart, VC-3, produced most of the tactical knowledge and a bevy of the tactical leaders of following generations of naval aviators. Despite all the perilous environmental difficulties and the oddities of both the radar systems and the aircraft provided, these pilots and their crews were some of the very few naval "futurists" of their time. They *knew* what the problems of tomorrow would be, and they *knew* how those problems should be attacked in search of solutions which *had* to be found. In their own time, and in their own way, and without ever voicing their feelings in such terms, they fully understood that for America to do its job in coming years, the U.S. Navy had to do its job, and for the U.S. Navy to do its job, it *had* to be militarily capable at any time, in any place, and in any weather. In large measure, the professional prowess of the F-14 Tomcat and F/A-18 Hornet aircrews of the 1990s directly reflected the foresight, example, and professionalism of the nightfighters of the Korean War days.

So in the fall of 1952 I found myself in this anachronism known as VC-4. I came from a diverse background that included tours as a "blackshoe" on board a ship during the final year of World War II, a slow tour through flight training in the immediate postwar era, two years in a fighter squadron flying the tiny Grumman F8F Bearcat, and a tour as an instructor and a legal officer at a remote auxiliary air station in northwest Florida. My pleas to get back to squadron life had been ignored through 1950 and 1951. But by mid-1952, I was with the crazy nightfighters in VC-4, after the obligatory four months in training at Key West, Florida.

Changes, unusual moves, and learning to expect the unexpected are a way of all navy life, but there was a touch of lunacy to VC-4 that soon had me reeling. Unlike "normal" squadrons, VC-4 flew around the clock, in almost any kind of weather short of a hurricane, and no party room was big enough to get everyone together at one time. In a normal squadron, living was close. You got to know your fellow squadronmates quite familiarly. VC-4 had officers aboard who had never laid eyes on one another, much less flown or lived together. Most squadrons were close families, tight-knit and self-supporting. VC-4 had all sorts of cliques of friends and enemies. When a normal squadron went to sea, everyone went together. VC-4 had officers and men who seemed forever deployed, while others remained forever at home, the latter earning a significant measure of disrespect as the "Palace Guard."

It soon became obvious to me that the det OinCs were the real bread-and-butter operators of the whole crazy business. They had to fly, teach, administer, and manage just like a normal squadron CO, but they had precious little help in doing the jobs. After forming up the det at Atlantic City, the OinC was the one most concerned about all the preparatory training of both officers and enlisted men. The deployment date of a carrier was the hard deadline that simply *had* to be met, irrespective of people, airplane, or weather factors. And any smart OinC knew that once finished and sent off to join the air group of the carrier, he was pretty much on his own. So, in general, VC-4 dets were small bands of half-forgotten men led by a very young, very worried OinC. Yet, ironically, more was expected of them than of any of the other carrier fliers.

I later learned that, with time aboard ship, and with reasonably intelligent CAGs and ship COs, things usually changed for the better. Because the VC-4 det was flying in conditions too severe for the others, respect for the nightfighters rapidly grew among the rest of the air group pilots. With success, they often became the elite of the elite. And because you had to be a little flaky to even want to fly off carriers, and absolutely mad to do it at night, VC-4 pilots were forgiven for almost any act of aberrant behavior ashore. Because their OinC boss was very junior, there was little formality to the det's organization and little fear of any disciplinary reprisals. As a matter of custom and tradition, nightfighters were supposed to be a bunch of crazy, suicidal bastards.

Within a week or two of my arrival in VC-4, I was given a fast verbal checkout for the prop F4U-5N Corsair, issued all the necessary gear, and turned loose to dash around the afternoon Jersey skies. This initial checkout was soon followed by an intensive cram course in how to use the Corsair at night. Formal classes were held during daylight, and the flying was almost always in the dark. In general, there were three different courses—night intercepts, night bombing and strafing, and getting on and off a carrier.

The intercepts were truly great fun, but kind of wild. VC-4 actually owned a large ground radar fighter control facility, complete with soft blue neon lighting, a large room filled with radar scopes, radios, status boards, and most of the electronic wizardry of that era, all housed in a shiny new concrete and steel blockhouse topped with a huge, slowly rotating radar antenna, just across the parking lot from the ancient wooden hangar. Although it had some official name, we knew it always as simply "the Blue Room."

CAPTAIN GERALD G. O'ROURKE

Most pilots in Composite Squadron 4 at Atlantic City in the early 1950s honed their nightfighter skills in the venerable Vought F4U-5N Corsair in the skies off the New Jersey coast. (Courtesy National Air and Space Museum)

The facility was run by a fat, weather-beaten old lieutenant commander named Joe Ross Riggs. Joe Ross was neither fish nor fowl. He wore wings but I never remember him flying. He was far too old for only two and a half stripes, so was obviously a longtime passover. His uniforms said the same thing—grungy, frayed, and threadbare, shiny pants seat, coffee-stained tie, open-at-the-collar shirts, and a greenish tinge to the gold stripes and brass buttons. At a squadron inspection, or even in a training lecture in the Ready Room, Joe Ross came across pretty poorly. In the radar facility about midnight, he was the ultimate professional, unsurpassed in his ability to conduct radar intercepts.

Here was pure bedlam—but organized and controlled, right down to the very smallest detail. The big radar atop the blockhouse rotated steadily, "painting" blips of every aircraft within a hundred miles or more of the station. All of the numerous intercept training flights were tracked, controlled, and monitored by this radar. Pairs of fighters would be under control, most sharing a chunk of sky off the Jersey shore. Each pair would be rigidly directed by a young officer or enlisted technician seated before a huge radar display scope. The grease pencils were flashing everywhere. All radio transmissions, incoming and outgoing, were broadcast out of hidden speakers as well as through the earphones on each controller. There was a lot of swearing and fuming and fussing on the part of the controllers, but only to the room—never on the air. On the air was always a cool, crisp set of orders and information barked out in staccato fashion to the pilots aloft.

At each controller's elbows were two sailors. One kept records, furiously scratching down the results of the training examination which was always in progress. The other was the gopher—the guy who gets coffee, soft drinks, cigarettes, new grease pencils, or rags, and who runs to the central control desk to carry messages to and from the single man who knew everything going on, every second, among all pilots, all aircraft, all controllers, in all weather using all kinds of equipment: Joe Ross Riggs!

Like the pilots aloft, the controllers were themselves students. Nonflyers, these men were temporarily assigned to VC-4 from their carrier, cruiser, or destroyer, sent to Atlantic City to learn the fighter control business from the navy's master craftsman. They learned much more than basics from Joe Ross. He cared little about what made a radar work, or test equipment or radios. He cared immensely about using the equipment to master the invisible night. He cared not one whit for rank, or privileges, or for the inept and uncertain. His motto was that of the true "operator"—"Do something—right or wrong—but do something!"

While he and his troops provided the ground-based teacher corps, the actual airborne training was a highly personalized situation. Each pilot, in general, trained himself. A small group of pilots who had returned alive from a previous det deployment were supposed to provide the corps of instructors. In fact, primarily because of the unexpected demands for replacement pilots in deployed dets, the "instructor corps" was constantly undergoing personnel changes, special assignments, or ferrying aircraft around the world. As each newly arriving pilot was assigned to a det in the formative stages, the OinC of that det became responsible for most of the training. As might be expected, there was no real training syllabus, no agreement on how much flying or schooling each pilot required, and very little standardization of operating procedures, excepting in the velvet-fisted domain of Joe Ross Riggs. The ultimate results, of course, were spotty. One det would be superbly prepared, the next only hastily exposed to what lay ahead. Worst of all, there were far too many stupid accidents, many of which were fatal.

Until I had been around for awhile, this grim aspect of the business wasn't apparent. All naval aviators are routinely exposed to, or involved in, aircraft accidents. That's accepted as almost a hazard of the trade. In carrier work, where dangers abound, accidents tend to be more frequent. In the night carrier operations of those days, accidents were so frequent that they were considered commonplace and unexceptional. Whenever a det departed, the aircraft they flew off in were more or less written off. No one expected that all of them would ever come back to Atlantic City. Perhaps one or two of the four, occasionally three, but never all four! Unfortunately, the same negativism tended to extend to pilots as well, whose return-safe rate wasn't much better than that of the aircraft. Between the pilots lost, the pilots maimed, and the pilots who decided to throw in their wings, precious few dets ever returned with the same resources they took with them.

After I had learned more of the details of the business, and had begun to understand the strange rules of the game, I was forced to do some serious thinking. After a few narrow escapes, the realization grew that most of the accidents were avoidable and that the pilots were themselves only partially at fault in any single one of them. Others invariably had a hidden finger in the pie—the weathermen who gave poor predictions, the ground crews who hadn't learned to be observant enough, the air traffic controllers who were drinking coffee when they should have been concentrating on a scope, or—far too often—the senior naval commander who didn't understand the night business, but who wasn't the least bit reticent to order planes and pilots aloft and put them into situations well beyond the limited capabilities of a young, inexperienced pilot.

After full realization had dawned upon a good VC-4 pilot—presuming, of course, that he was lucky enough to still be alive and unlucky enough to be stuck in the business—the natural, human will to survive was given full rein. Limitless cynicism took over. He began to use, but not depend on, all others. He would constantly remind himself that once into a darkened cockpit, it was "me against the world"—from the plane captain through the deck handlers, the catapult crews, the air controllers, the Landing Signal Officers, anyone else who could call him on a radio, and most of all, the old senior officers who commanded the carriers or directed the flight deck, or the combat information center. If he kept these people all in mind as grave natural enemies bent upon personally killing him, he had a fair shot at short-term survival. Get suckered in for even a moment, in any way, and he was dead—and the records would show that it was all his own fault anyway!

Within a few months, I began to understand the underlying purpose for the nightfighter program which the squadron sustained within the Atlantic Fleet. Only a handful of naval aviators appreciated the importance of its continuity in the early 1950s. In those years, the residue of military hardware and tactical thinking of World War II were very much in evidence everywhere. The critical need for night and all-weather capabilities had been demonstrated in the last years of the navy's great war in the Pacific. A dozen or more programs had been started in those days, each of which would have contributed greatly, had the war continued. When it stopped, when the great economy waves rolled over naval aviation, and when most of the experienced flying talent returned to civilian life, all these programs were abruptly curtailed, and whatever hardware remained was used in its half-finished state.

Even though the diehard nightfighters, such as our VC-4 commanding officer, Capt. Joe Gardner, felt that the night program was getting the short end of the stick, the official navy viewpoint was that there was no other recourse. The money wasn't there; the people weren't there. Nightfighting was, after all, only a small segment of a carrier's total mission. Furthermore, it was a dangerous business, and a costly one. The airplanes were very expensive, and they invariably had far more accidents. So the night and all-weather requirements were more or less ignored in the late 1940s. As the Korean War developed, the program took another sharp turn. Nightfighters became real people aboard navy carriers, not just some crazy bastards who interrupted the movie schedules on the hangar deck. A few brave day squadrons actually flew some night attack missions from several carriers, despite the shortcomings in their equipment and their training.

For the large composite squadrons back in the United States, the new demands for additional all-weather forces were hard to meet. Pilot training was a major problem. Adequately prepared numbers were just not available. Key West and Hawaii were pumping up their output of basic night training, but the big VCs were simply not organized to follow up with the finishing courses, the carrier qualifications, and the formation of the new detachments.

This was the atmosphere in VC-4 when I arrived. The squadron was literally loaded with pilots, but desperately short on experienced nightfighters. Of all the airplanes in the large assortment within the squadron, I fell in love with the worst-looking one of the bunch—the F3D Skyknight. The Skyknight, unfortunately, was practically phased out of naval aviation before it ever got in! Because there was no existing program for training the radar operator crewmen, and because the plane was big and slow, and the radar equipment unusually complex, a decision was made to send all the F3Ds to the Marines for land-based use, where it fit in very neatly as a modern jet replacement for their aging, two-place, prop-driven F7F Tigercats. Somewhere along the line, in some remote corner of the Pentagon, there was an inadvertent lapse in the great naval bureaucracy and a small handful of them had been assigned to VC-4!

Ever since early in the war, nightfighters had been airplanes originally designed to do a daytime job, then modified with radar and instruments to work at night. The F3D was started with only nightfighting in mind, right from the very beginning. It had good stability for instrument flying, a lot of radar, two engines for dependability, two men aboard to do the work, and plenty of fuel to stay airborne for long periods. It wasn't pretty and it wasn't graceful. The nose was large and fat. The fuselage seemed to hang from the high wing. The two engines made ungainly bulges on either side of the lower fuselage, like a whale cow with simultaneous twin pregnancies. Just aft of the straight wings there were two large panels—speed brakes—on either side of the fuselage which could be extended to slow the airplane down. The tail was very conventional, just like the tails of a prop plane.

The F3D carried a crew of two in side-by-side seating and had one of the most complex, sophisticated radar systems known to man in that era. There were three separate radars and four different radar scopes in the huge cockpit! The radar system's design required a well-trained operator, yet the navy had no program to train such people. Half of the already small group of nightfighter pilots in the navy felt that a single-place plane was a better design anyway, so the planning was always mired in philosophic arguments and postponed decisions. The complexities of the radar system demanded the best available electronics maintenance talent—petty officers who were in critically short supply both in the navy and in civilian life.

When the Skyknights arrived in VC-4 in 1951, their program was assigned to two pioneering young officers, McCrary and Null, who strongly supported the two-man concept. They readied the first pilots, mustered up an assortment of radar operators from all walks of the navy, conquered several serious engine problems, and actually got two four-plane detachments onto carriers and into the Mediterranean for deployments in 1952. Lieutenant Commander Null was tragically lost in a crash while so deployed.

In VC-4, there was always a "new boy" checking out in one airplane or another. I got my initial jet ride in one of the newest Big Banshees. The sensations were truly weird. Wings were well aft of the cockpit, so far back that you had to turn your head to catch a view of a tip. I felt naked and unprotected sitting way up front, without the prop and huge engine blocking my view. I wasn't accustomed to having two throttles, and felt like a real king with all that power at my fingertips. The cockpit smelled new. The controls of flaps and gear and speed brakes were electric, requiring only a flick of a switch instead of the older manhandling of a hydraulic control. Although the tail controls were still connected to the stick by cables, as in the older planes, there were various boost systems and a changing "feel" of the stick dependent upon the plane's airspeed. This was strange and scary. Airspeed would build up at a fantastic rate. There was very little noise, save the crackling of the radio and the gushing of the plane's ventilation and pressurization system.

In late 1952, VC-4 received word of a commitment to supply a detachment for a carrier deploying to the Pacific, for Korean War duty. There was much soul-searching about what type of airplane should be sent. The Skyknight had been requested by the Pacific Fleet, but its limited experience in the Mediterranean was hardly enough to certify its success for war. The new Banshees were in very short supply and had tail troubles. Corsairs were plentiful, but this was a backward step in progress. The messages flew back and forth, with proposals, counterproposals, decisions, and modifications to those decisions.

All through this period, we who had volunteered to go on the deployment were kept in constant turmoil. Plans would be made on one basis one day, only to change the next. All of us were Corsair-qualified, but were eager to get into the jets as soon as we could. As the message traffic went flashing around the world, we frantically checked out in one plane, only to find a day later that we were going in another. My checkout ride in the Skyknight was delayed for several days by a storm that blanketed Atlantic City with a wet and clinging fog. Each day's delay robbed me of that much more flight time in which I had to master this challenging airplane. I grew quite bitter about the whole subject of weather, and about Atlantic City weather in particular.

In no time at all, we found that daylight flights in the F3Ds were wasteful. The pilots could get some good training in basics, and with the normal lousy winter weather, there was plenty of opportunity for very realistic flying on instruments. However, in daylight, the radar operators were forced to bury their heads in the radar shields in order to keep light off the tubes. With the shields in place and the ROs buried therein, the pilots, who were the teachers, couldn't help very much. In full darkness, the shields could be stowed, all four scopes were easily visible, and the pilot could lean across the center pedestal and supervise the training very nicely. Since our mission was nightfighting, we lived the night schedule as best we could, coming to work around noon, getting into the air at sundown, and getting back home well after midnight.

We were training as pure interceptors who work in either darkness or clouds, or both. Our ultimate job was to find another airplane in this airy ink, and to either shoot it down or close for a visual identification. This basic mission had a dozen variants, such as night patrols, alerts, fighter sweeps, etc., but the essential ingredients were the same—leave home, fly in and through the worst of nature's elements, make intercepts, and get back home in one piece.

We had to first become the best instrument fliers in the world. You can practice instrument flying in a lot of ways—in a Link trainer, or in fancier modern simulation devices, or in a training plane under a canvas bag on a clear day, with a safety pilot in the other seat, or in a single-place service plane with a chase pilot flying close aboard in another plane. It goes without saying that the best instrument training is done at night. Start a young pilot on a clear, moonlit night, and work him up to the black ones. Fly him over water or vast western wastelands, where lights are weak and few. Let him learn what instruments can do for him, and he'll soon start making them part of his flying nature. Like all naval aviators, we had supposedly been taught instrument flying while earning our wings years before at Pensacola. In fact, we had only recently learned how to really use the gauges in the night sky over the Florida Keys.

A nightfighter's primary need is to master the hostile elements of nature. He's got to be smart about this. He's got to respect weather. He's got to learn what can and, even more important, what cannot be done in weather. Most of its hazards can be overcome. Those that can't must be avoided. Because weather becomes his most intimate bedfellow, he soon becomes a fairly proficient forecaster. Unlike the airline pilot, whose relatively simple task is to safely fly through some clouds now and again, the nightfighter must learn to use weather conditions to his tactical advantage. Clouds often make excellent hiding places from which traps can be sprung. Good weather at his "Home Plate" and poor weather at his enemy's means more fuel available to expend in a dogfight. Split-Ss and tight turns at near stall airspeeds must be made on the "feel" of instruments, not seats of pants!

Next in importance is his radar. It is his eyes in the night. Like human eyes, radar can see some things, but not all things. Radar can be fooled. It can look in the wrong place, it can't see behind a wall, it only sees in a pencil beam pattern, so it has to move often, and like some animals and middle-age humans, it can see well afar while almost blind to targets close at hand. Once a pilot has mastered the art of flying by instruments alone, he is then ready for introduction into the mystic art of airborne radar.

In VC-4, where almost all of the flying was solo, we had our full share of both artists and klutzes on radars. Surprisingly, the artist was not always that good at instrument flying, while a real radar klutz often proved to be a superb master of night and all-weather flying. In fact, a single pilot truly great at both was rare. We had several of these priceless rarities, and even a couple who were naturally good teachers as well. The admiration and respect shown to them was truly awesome, and had absolutely nothing to do with rank. The printed nightly flight schedule automatically listed them as flight leaders/briefers any time they flew, totally irrespective of the makeup or the rank of the rest of that flight. And whenever a good heated discussion would get going in the Ready Room, their opinion would be solicited and their decision accepted as gospel. One of these, John Wissler, was only a lieutenant (jg), nearly the junior-most pilot of the whole outfit! Years later, he was also one of the navy's best flag officers.

The most basic part of the radar intercept art is to search the sky and detect a target. Unlike human eyes, which cover a broad area in a single glance, radar eyes look only within a small cone, much like the beam from a narrowly focused flashlight. To provide the panoramic view needed for searching, the interceptor's radar beam is moved left and right, up and down, in a pattern, "looking" about 60 to 65 degrees to each side of the nose.

One good reason why the Skyknight was called a Whale was because its antenna was huge. As our familiarity with the plane increased, we became more and more attuned to the movements and the idiosyncrasies of our massive radar eyeball, which scanned the skies from its perch directly in front of our cockpit. We grew familiar with its growls and whines as it swept left and right. We could feel its angry thumping as it reversed direction at the end of each sweep. At high scanning rates, it could actually make the plane fishtail through the sky in tiny undulating lateral waves that fed back to nudge the feet on the rudder pedals.

Our normal hops were two-plane affairs, and were guided from the radar station at the field. There one of Joe Ross Riggs's controllers ordered us to fly this way and that way, until we two were separated by many miles. He then turned us, announcing one plane as the "bogey," or simulated enemy attacker, and the other as the "friendly," or defending fighter. As we closed, the friendly searched the skies with his airborne radar, picked up the target, and took control of his own intercept from that point on. After carefully maneuvering around to a point just a few hundred yards astern of the bogey, the intercept was complete, and the ground controller would once again take control and separate the planes for another run.

A critical portion of all VC-4 training was occupied by "bounce hops." These were periods of practice landings on a runway which were designed to simulate, insofar as possible, actual carrier operations at sea. In daylight, bounce hops are interesting at first, boring shortly thereafter. At night, bounce hops are terrifying at first, fascinating forever. Rubbernecking from the LSO's "platform" on the runway was encouraged for neophyte night-fighters. A half-mile stroll across a few runways and grassy infields in the evening breeze brought you to the huddle of shadows surrounding a noisy radio jeep parked in the grass. Once there, the outline of the simulated carrier deck became visible. The normal lights outlining the runway were off. Small flashlights were laid in a rectangle the size of the carrier's landing area. The LSO, bedecked in a flight suit wired with small Christmas tree lights, stood to the port side of the outlined carrier deck. His flags or "paddles" were also outlined with small lights.

A pilot flying a carrier approach could pick up the LSO visually as the airplane completed about half of the 180-degree turn in the approach. For the LSO, the dim silhouette of the plane had almost to be imagined. Multicolored approach lights told the LSO what the plane's flight attitude was—too slow was red, on speed was amber, too fast was green. Small changes in throttle settings by the pilot in the groove were detectable through ever so slight changes of pitch in their whine. The LSO stood with arms outstretched, waving his signals, the pilot responding, until finally the LSO slashed his right arm across his chest as the plane came abreast of him. Power was cut, a huge black outline rushed past, red wing light seemingly an arm's length away. There followed a screech of rubber on concrete, engines wailing under maximum acceleration, and the plane bobbed back into the air to set up for another pattern.

It all took place as fast as you read it. The LSO turned to his helper, muttering comments while he scribbled in a dog-eared green pocket ledger. The process was repeated over and over, eight to ten passes per fuel load, followed by a final landing, a quick refueling, then out again for more passes, then finally to the Ready Room for debriefing.

When I first watched the Whales work at night, I couldn't imagine how anyone could master all the complexities of something so large. I was astounded to see the big birds whooshing around the black night, turning almost as easily as props, and slamming on and off the runway. The colorful lights against the black and gray horizon, the foreboding darkness of the heights of trees around the field, the busy professionalism of the radio chatter—zip, zip, zip—there was something happening every second. This was my challenge. I felt like a fifteen-year-old schoolboy, in love with the Whales and with the night.

Lt. Comdr. Howard Leroy Terry was our appointed team leader—the officer-in-charge of the detachment, and the man most responsible for our success or failure. Five other crews filled out the flight personnel, and some fifty-one ground technicians and mechanics completed the group. Terry was the oldest of the officers, easily the nicest guy in this world, and a truly dedicated career naval officer. He had graduated from the Naval Academy in the class of 1942, served aboard ships during World War II, won his wings at Pensacola, and had gone through a postgraduate course in engineering. In

VC-4 he was a fairly senior lieutenant commander, but because of the time spent in study, his actual flying experience was not commensurate with his rank. Nevertheless, he was a fine leader, with a serious sense of purpose, a grand sense of humor, and a deep sense of responsibility for people. He felt that all personnel should and could be managed on a personal basis, and he did just that in forming the team. We were all volunteers—from Terry down to the lowest airman.

For a wide variety of reasons, we all wanted to get into the Korean War before it folded up. I was primarily motivated by professionalism. I had enviously followed the combat successes of many Naval Academy classmates from the secure, hateful safety of my legal office on a training command field in Florida. I wanted a piece of the action, and this was my chance.

It happened on a typically nasty New Jersey early spring night—18 March 1953—only a few weeks before our departure. The runway was slick and shiny, shimmering under the marker lights, when they took off. The rain had been hard a few minutes earlier, when I had returned from my flight. Their Skyknight was fully loaded with full wing tanks, as a precaution against the vagaries of the midnight fog and rain. About 11:00 P.M. their plane hurtled down the runway; they couldn't have been airborne more than two minutes when it all came to an end. The plane cut a swath through the scrub pines and swamp oaks less than a mile from the base, narrowly missing several houses, leaving a quarter-mile streak of flames in the cold, wet night.

The accident was exhaustively investigated, as are all aviation tragedies, and a probable cause was formulated. Years later, with the retrospect of experience, I know that the officially stated cause was wrong. Most likely, Terry "lost the bubble" for a few seconds after takeoff and simply flew into the ground. Maybe it was a bad gyro horizon, maybe ice formation in instrument sensors, maybe a distraction in the cockpit. Whatever, it cost the lives of two of the world's finest people—Terry and his radar operator, O'Neil.

A few hours later, I was told that I was now the OinC of the det, and that we would go with five aircrews, not the six we had planned for. There just wasn't anyone else ready for that job at that time. I viewed the selection with the growing cynicism of the professional nightfighter. The trite expression "mixed emotions" was particularly applicable. I was glad to get my first command, scared to prove unworthy to the task, derisive of the honor as the only available choice, deeply worried about the team prospects after the night's tragedy, and overwhelmingly sorrowed by the loss of two fine shipmates.

The night of horror, even at 4:00 A.M., was not yet over. At home, I found the lights ablaze, a tearful wife, and our daughters' beloved puppy shaking and quivering with spasmodic loss of nerve control. This was the culmination of a week of trips to the vet, odd medications, dewy-eyed girls, and shaking heads. So, at 4:00 A.M., there was yet a further chore, before the girls awoke to heighten the crisis. I drove to the outskirts of town, along a dirt road, did away with the puppy, and buried him in the cold, soggy marsh. As I did, I could not wipe that other scene of rain, mud, scrub pines, and death from my mind.

Instead of shattering the detachment, the loss of Terry and O'Neil solidified us. We were rudely shown how close we flew to death, and how great a task lay ahead. We had yet to carrier-qualify the team, an evolution that took place on board a ship off Jacksonville later that same month. In preparation, we bounced interminably on the runways at Atlantic City. A regular LSO from the squadron supervised this training. In the nightfighter world, there are many strongly individual characters, but the most independent of all are the LSOs. VC-4's Paddles prepared us well. We flew down to Jacksonville one evening, hopped aboard ship in the Gulf Stream the following morning, day-qualified that same day, night-qualified the following evening, and flew back to Atlantic City in the wee hours of the next morning. With about fifteen day landings and eight night "traps" behind us, we began to grow more confident about the future.

Those few minutes—ten to fifteen at most—immediately subsequent to a good night carrier landing in very poor weather are gloriously exhilarating ones in a pilot's life. He is almost bursting with sensations, with emotions, with enthusiasm and excitement and a sense of personal achievement. All fears, worries, and self-doubts are gone! Nothing else in life—no future challenge of any kind—need be avoided! He can do anything! But the twenty or so minutes leading up to that "good" night carrier landing frequently were what made the afterglow so, enjoyable. They would go something like this:

At the conclusion of a practice intercept mission, during which you had usually been under intermittent radar control at ranges of twenty-five to fifty miles from the ship, you had to make an assigned overhead time. The ship's radar couldn't help with vectors within about fifteen to twenty miles, and often not beyond. Sometimes they couldn't see you at all, and if you were transferred over to control of a destroyer's CIC, you were lost from start to finish. Remember, too, there was no TACAN in those days! Aside from your own radar, your only navigational aid was a crazy low-frequency thing called YE/YG, which broadcast a different Morse code letter in each 30-degree sector around the ship. Assuming the YG was in synch with real-world geography (it often wasn't), and assuming no thunderstorms were around to render it useless, you would track the signal inbound, at say 4,000 feet, until you got a rapid change to another letter. Now you were overhead.

When directed, you would fly outbound, off the ship's port quarter, for about five minutes, then make a descending teardrop turn and shoot for coming up the starboard side at 500 or 1,000 feet. *IF* the ship was into the wind, and *IF* the plane guard destroyer was in place directly astern, and *IF* your radar was working okay, and *IF* you hadn't tried to "sneak a peek" too often, you'd be in good shape, at 150 knots, just passing the starboard bow, on the landing heading. You'd continue upwind for a couple of minutes, then turn, slow up, put landing gear, flaps, and hook down, and descend, shooting for a position abeam, landing check completed, hanging on the power, at 125 feet and 100 knots.

Here was where you had to fight the overwhelming temptation to "sneak a peek" over the port side—a deadly vertigo-inducer. A good LSO could help here with a terse remark about being high, low, too far abeam, looking good, keep it coming, etc. Somewhere about then, depending most upon your best estimate of the wind, and your confidence in the ship to be on its claimed heading, you'd start the all-important final turn, still on the gauges. With about 90 degrees of turn to go, maybe after a couple of quick peeks, you'd come fully off the instruments and start trying to see the LSO and his paddles. From there on, it was a piece of cake—as long as you didn't angle, overshoot, get too fast, get too slow, ease your bank, wrap up your turn, climb, descend, ease power, put on too much power, or get long in the groove. The turn had to be right on, allowing you to roll level right close in. If you had started high, or low, you'd be fighting speed all the way. In the Corsair, with enough experience, you could cheat a little by hanging in a lot of right rudder, left stick, and power, to keep Paddles in sight in a long-in-groove situation, but then, when you straightened yourself out just prior to the cut, you'd pick up speed like a shot. For a good pass, you had to stay pretty well wrapped up to keep sight of Paddles, then almost simultaneously roll out, take a cut, dive into the black pool, and horse back on the stick.

Waveoffs were frequent, often because the ship was still hunting the wind or otherwise unprepared. Late waveoffs were scary, since you'd have to go to full bore, clean up, dodge the ship, and get back on the gauges all at the same time. The AD Skyraiders, with a high cockpit and a short nose, were by far the best of the prop bunch, the "Hose-Nosed" F4U Corsairs clearly the worst. Almost any overshoot in a Corsair brought an immediate waveoff from a good LSO, since he knew you would either try the crab bit or lose sight of him trying to correct.

The props were bad enough, but with the new jets, without benefit of any CCA, landing mirror, divert field, or airborne tanker, it was Russian roulette all the way to touchdown. Initially, we tried to make the approach in the same way we had been in props, but using a higher and farther abeam position. Angle-of-attack gauges were new and unreliable, and it was almost impossible to hang on the power in the turns, so speed control was tricky. Seeing Paddles, of course, was much easier, but sort of useless, because you needed more time for corrections. In the F3D, you had to learn to concentrate on one of the several LSOs appearing through the various windshield panels. Lineup was the critical item, and you had to do all of that by yourself. A good destroyer driver could help by holding his station directly astern of the carrier. And the centerline deck lights of those days were notoriously unreliable. In fact, the whole ship lighting setup was usually pretty awful, and invariably became a very sore bone of contention between the ships and the night guys. Carrier night lighting had originally been designed more to confuse World War II submarines than to help aircraft get aboard, had been subjected to piecemeal improvements, and, in general, was treated with humorous disdain by flight deck officers.

Any carrier landing is far from gentle. The plane actually impacts the deck, which itself may be rolling or heaving up and down in a good sea. Rather than a landing, it is more like a "controlled crash." The plane's tail hook almost simultaneously snags a cross-deck arresting wire, initiating a fierce, violent deceleration of about 100 knots of relative speed in less than three seconds. With the wire is snagged, a "trap" has taken place. An enormous amount of energy is dissipated into the innards of the ship in a screeching of wires over cables, hydraulic fluid screaming past tiny valve orifices and air grunting under the compression of monstrous pistons. If, by some strange quirk of happenstance, the hook does not snag a wire, the plane hurtles forward into several wire barriers, and quite often further on into "the pack" of other aircraft parked on the forward half of the flight deck. Such events are truly disasters, with a death toll of at least a half-dozen, sometimes many more than that.

But that was the olden days, in the early 1950s, before the arrival of the angled deck and mirror landing system, which have made dramatic improvement to recovery operations. With the angled deck, the landing plane can skip over all the wires and merely take off for another try. As a safety measure, at the moment of touchdown, instead of taking a "cut" of power, the pilot actually jams the throttles full forward to maximum engine thrust, thereby ensuring enough power to climb out should the tail hook not snag a wire. This is called a "bolter"; while dramatic and awesome, it's actually a very straightforward aerodynamic experience, closely akin to the "bounce" in the bounce hops used for practice ashore, or to the "touch-and-go" universally used in training all pilots in field landings.

A carrier pilot never knows, until that violent deceleration takes hold, that the flight has ended. He cannot—must not—predict that the trap will actually take place. To do so inevitably implies slow reaction, late power application, and a very real potential for "dribbling" off the angled deck with insufficient flying speed after an unexpected "bolter." The mirror landing system, another British idea, started as a large mirror that reflected a light source for pilot guidance on the glide path, but is now actually a system of Fresnel lenses that accomplishes the same purpose.

In olden days or modern times, the sudden change, for the pilot, remains much the same. At one second he is superbly alert, flying with supercritical sensitivity and ultraresponsive control of a powerful, but fragile, airplane. The next moment he is merely riding along on a furiously descending wave of dissipating energy. An old Ready Room description of a carrier landing, oft-repeated in recruiting speeches, is "The most fun you can have in life with your clothes on!" There is a sharp relief from tension, a quick flight from anxiety, a burst of pride of performance, a deep satisfaction of skill, a tremendous surge in personal maturity and responsibility, and a deeply rooted magnification of confidence in self.

Alone, much later, you think back to the experience, remembering again a thousand or more infinitesimal details. You grimace a bit at the thought of your initial speed when you started the final descent. It was too slow. Only you, and perhaps your RO, know that. "Must remember to watch that more closely the next time. We could have busted our asses right there." Your speed control was good thereafter, with throttles almost motionless, but you say to yourself, "That air was really steady. No turbulence to speak of even in the burble just aft of the fantail. How bad would that approach have been in rough air?" And the intercepts, the myriad nuances of the radar work, the feel of the hard turns on instruments . . . on and on.

This is a vital phase of the continuous learning process. It is the digestive period, when all the minute inputs are sorted, categorized, analyzed, routed, and embedded into the memory, where they must be ready for instant recall, reuse, in subsequent situations. After a few dozen such experiences, over a period of several months, you are surprised to realize that very little has changed in your inward thought processes. Each new landing is a new and demanding challenge. Each draws its fair share of adrenalin. Each has its merits, its rewards, its perils, its teaching, and its own peculiar idiosyncracies. Any of them can ruin your whole day—or life.

Outwardly, you have joined the first line of the professionals. You can now feel very comfortable in teaching others. You can assess how others are doing at the same chore, without even watching them fly. Simple conversations in the Ready Room are sufficient. Now you have developed enough self-confidence to lead with conviction. You know what your limits are, and what extremes of weather and sea you can master. You also know the ones you can't. You have a very good handle on the limits of the other pilots and those with whom you fly. Most important of all, you can now "converse with conviction" with the carrier skipper, the air operations officer, the air wing commander, the air officer, and with the admiral and his key staff members. Now, and only now, can you stand up to any of them on the all-important questions of when and how the carrier's all-weather flying should be done.

Now, you're a nightfighter.

4
Flying with the Blues

Captain Arthur R. Hawkins

Arthur Ray Hawkins was born in Zavalla, Texas, on 12 December 1922, son of Alva M. and Gillie B. (Russell) Hawkins. He attended Lon Morris College in Jacksonville, Texas, had cadet training at the naval air stations in Dallas and Corpus Christi, and was designated a naval aviator and commissioned ensign 1 January 1943. Subsequent service included extensive operational and staff duties ashore and afloat, and he advanced in rank to captain, to date from 1 July 1963. He was placed on the retired list effective 30 June 1973.

During World War II, Captain Hawkins served as navigator and gunnery officer of Fighting Squadron 31 based on the USS Cabot *(CVL-28). He participated in all naval engagements from the Marshall Islands operations to the fall of the Japanese Empire, and was awarded the Navy Cross three times for extraordinary heroism in aerial combat against the enemy. He was credited with shooting down fourteen enemy aircraft, sinking various enemy ships, and assisting in the sinking of the Japanese battleship* Ise.

Captain Hawkins's operational and staff assignments in the postwar years included command of the U.S. Navy's Flight Demonstration Team, the Blue Angels, Attack Squadron 46, Carrier Air Group 1, USS Caloosahatchee *(AO-98), and NAS Atsugi, Japan.*

CAPTAIN ARTHUR R. HAWKINS

Comdr. Arthur R. Hawkins, former commanding officer, U.S. Navy Flight Demonstration Team, the Blue Angels, 1951–54.

Captain Hawkins's numerous decorations include the Navy Cross with two Gold Stars, the Distinguished Flying Cross with two Gold Stars, and the Air Medal with three Gold Stars.

After my discharge from the navy in October 1945 I went back to college, the University of Cincinnati at that time. I got word that I had made regular navy and in November 1946 I started flying for the navy again, at Jacksonville, Florida, flying OS2U and SC seaplanes. I didn't make the transition then from fighters to scout planes; the navy did. Actually, each battleship and cruiser had an aviation unit on board, and so there were usually four aviators on board ship—a senior aviator and three junior officers plus a crew of about twenty-two who maintained the two airplanes—the battleships carried three, but the cruisers carried two usually.

After the war was over, the brownshoes (aviators) and the blackshoes (nonaviators) didn't get along too well, I guess, and they just got fed up. All the senior aviators got out. They'd had it, and it ended up they had no senior aviators; so they took fourteen fighter pilots—lieutenants—made them senior aviators, and sent them through training to replace all these senior aviators they had lost. We ended up—fourteen of us—going through training at the same time, every one of us fighter pilots.

I was on the USS *Portsmouth* (CL-102), a light cruiser. She was a flag ship for commander, Naval Forces, Mediterranean. All the flying on board ship was done in SC-I Seahawks. The SC was a single-seater with no crewman, so the pilot had to make his own recovery. A recovery is made by the ship turning into the wind and creating a slick on the water inside the wind line. Then you come in and land on the slick water that the ship created by making a sharp turn. The ship drags a sled made out of rope or heavy metal wire; on the bottom of your float on the airplane you have a hook that catches onto the net as you taxi up. Once you're hung into that net, it's just pulling you along. So with the SC by yourself, you had to cut your engine because the crane then was going to come down to pick you up. You had to get out, hang a foot inside the cockpit and unsnap a panel up front so your sling would come out, and reach around and get your sling. This big crane is coming down, and you've got to hook it on yourself and jump back in the cockpit. Then they'd hoist you out.

Well, it wasn't uncommon on the rough seas that while you're trying to do all this you'd shake off of the sled and start sliding aft, so you had to hustle yourself back in the cockpit, restart the engine, taxi back up and pick up the sled again. So it wasn't too good an operation for one man, although it worked all right.

It was obvious that the use of the airplane in scouting and observation work was fading with the helicopter coming into service. I convinced them on our ship to take off one catapult. We could put one plane in the hangar and one on the cat and launch two almost as fast as we could with two cats. So they started out by taking off one cat, and then after that was done it was easy to convince them to take the other cat off; then you've got a perfect deck for a helo.

The SCs were there for scouting purposes, as a spotter for the main battery, and for search and rescue. You could pull your seat forward and slide somebody back in there and bring them back. But with the radar improving so much, it was getting to the point where spotters were not necessary. The radar could do better than a spotter could.

I did two cruises over to the Mediterranean, each one for six months. For a career naval aviator, that was excellent training. For me it couldn't have happened better, because I was able to get qualified OOD under way and get those tickets taken care of. Then, luckily enough, I left that and went right back into the type of flying that I enjoyed. In the spring of 1948, I got my orders from the cruiser back to the Instructor's Advanced Training Unit (IATU) in Jacksonville, the unit that trained instructors who went all over the training command. The navy's Flight Demonstration Team, the Blue Angels, were attached to IATU. It just so happened they were having a couple of pilots turning over at the time, and they approached me to join the team, I guess because of my war record.

So I joined them during that time, and we moved to Corpus Christi from Jacksonville right after that. The team operated out of Corpus with F8F Bearcats, piston-engined aircraft. Then the jets came out, and they wanted us to go into jets, so we went to the jet squadron out on the West Coast, VF-51, where they checked us out in jets. They had the Lockheed TO Shooting Star, later called the TV; and the Navy's FJ-1 Fury was the first jet they had, and Air Group 5 was the one that got them.

So we got checked out in the trainer, the TO, and then came back and picked up our Grumman F9F Panthers. It was a big change going from a prop plane to a jet—an awful lot as far as the tight formation flying we were doing. You're so used to fighting the rudder and prop torque and so forth. Once you had the jets, you had no rudder to speak of. It was there, but it didn't do much for you; and the acceleration was slow, whereas with an F8F you had an instantaneous acceleration. You got out of position and could get back in a hurry. So it was a lot, getting used to the jet—lead and lag type of thing, being able to anticipate if you were going to need a lot of power here in a minute. Then you didn't worry about the prop cutting somebody's tail off. All in all, it certainly turned out an easier airplane to fly formation in.

It wasn't as spectacular, because you couldn't do the things that you could do in the F8F. That F8F could do a show and stay right in front of the crowd and never leave the confines of the field, whereas with the jet you had to take it out and turn it around and give it a little leeway and spread the show out. With the F8F you could do stuff like a loop on takeoff, but you were certainly not going to do that with a jet—not those we had then, anyway.

We learned the routines for the air shows by starting out, two people, flying formation. You'd go out and just fly wing on one guy and do rolls, loops, nip-ups—nothing you hadn't done all your life in an airplane anyway. It was just a matter of, "Okay, we're going to formalize this thing and do something where each guy knows what he's doing." And then the four of us would go out with a leader and just practice. Of course the routine would be set up as you went along, as you put new maneuvers in, took maneuvers out, shooting for about a thirty-minute routine. That basically was it—just practice. Once you got the routine down, it was set for that show.

Grumman F9F Panthers flown by the Blue Angels, 1949.

Actually, career-wise, Blue Angel duty hurts—and it helps. Certainly it's like losing two years from an active duty outfit, doing the gunnery and bombing and so forth. However, the prestige of having been in it is certainly a plus. You can imagine going on the road and living out of suitcases; it's not the glamor that it's cracked up to be, really, but it's certainly enjoyable flying. You get the kind of applause and public recognition that is afforded to very few in the navy in that way, but you try to hold it down as much as you can, because there aren't many naval aviators who couldn't do the same thing if they were pulled in as a trainee to take over a spot.

They have qualifications now that you need to meet—so many hours in a jet and so forth that they require before you can even apply now. There are other things that you have to take into consideration—living close together, public relations—but it's still sort of a personal selection. You report on board, and the whole team gets to meet and greet you and find out what you are and who you are before they say, "Well, I think you can hack it." Flight-wise you don't know until you get a prospect in the air, but if he's gotten that far you know that he's got the capability of doing it. But some people just can't live in a close-knit group like that. You certainly don't want a man who's got a drinking problem or any kind of problem like that, because it reflects on the navy.

They're an official squadron now, although they don't train like a fleet squadron. During the Korean War, we were flying jets and they needed a jet squadron. So the Blue Angels became the nucleus of VF-191, which was already a squadron, but they weren't flying jets. In July 1950 we took all our expertise and our men, plunked them into the squadron, and did our checking out of our own pilots and crew. I was the operations officer. It ended up that each one of the Blues had his own division; he picked up three kids and formed a four-man unit. So it worked out well that way.

In 1951, after a combat tour in Korea, our ship, the *Princeton* (CV-37), returned to the States and they pulled two of us who had been on the Blue Angels before and sent us back to Corpus Christi to help reform the Blue Angels. So they took Pat Murphy and me—both of us had been with the team before when we formed the squadron—and we went back to Corpus Christi to help reform a team. We picked up new pilots, and as a matter of fact three of them came from VF-191. That way we knew whom we were getting and knew their capabilities.

So we reformed the Blue Angels in Corpus Christi. Lt. Comdr. Butch Voris, who had formed the original team back in Jacksonville in 1946, came down and helped reform the team and led the team for almost a year. Then I took over from him as commanding officer of the Blues. We were then flying the F9F-5 Panther. We had flown the -2s when we went into combat in VF-191, and the F9F-5, which is the next step of the F9 series, had a little better engine with some improvements. We flew those on up through 1953.

At the end of 1953 we were assigned the F9F-6, which was our first sweptwing fighter. They could fly faster than the speed of sound, and they were just much more advanced aircraft. They had electrical trim tabs on the stick; they had the new "flying tail." The elevator on a normal aircraft will go up and down when you move the control stick fore and aft. With a flying tail, you rigidly lock that elevator to the horizontal stabilizer and free the stabilizer so that the whole combination, called a "stabilator," moves, like a huge slab. Instead of working with just the small elevator, you're working with the whole stabilizer, which was hydraulically operated, like power steering. You pull back on the stick and the hydraulics make the whole stabilizer move. But you could freeze it and adjust to the regular elevator at low altitudes. All that did was give you a bigger bite of the air at high altitudes, which let you turn tighter and gave you more control of the airplane. It was well designed to start with, but not too well thought out.

When we picked up the first F9F-6 aircraft for the Blues, we got six of the first thirteen the factory built. We went up to Grumman on Long Island, picked them up, and started home with them. But I didn't make it home. I had a little problem at about 42,000 feet on the way back to Corpus Christi. The flying tail ran away on me, and we found out later what happened. It was hydraulically operated and it had a slip valve; when you pulled the stick back, it opened a little hole, and the pressure went in and slid this valve back and forth, making the stabilator move. Well, I developed a leak on the downside of this valve as we were going along, so it started nosing the airplane over a little—over and over and over and it was going into an outside loop at 42,000 feet. At the bottom of this outside loop, which was about 32,000 feet, I started redding-out, which is the opposite of black-out. With negative "gees" your blood goes to your head, and in a black-out the blood is being pulled away from your head with positive "gees." So I had to bail out of the thing.

The F9F-6 also had a new feature where the pilot could eject himself through the canopy, which had never been done before. We needed some way to get out of the airplane when you couldn't get rid of the canopy, because the normal procedure for ejecting from an airplane was first to blow the canopy off, arm your seat, and then eject. This was done with one handle below the canopy on the seat to blow off the canopy, and then you had to pull the face curtain over your face to eject yourself out of the airplane. Well, our squadron skipper, Johnny Magda, was killed in Korea, and we were almost sure that he couldn't get rid of his canopy; and if you couldn't get rid of the canopy, you couldn't arm your seat. He went in and hit the water still in the cockpit and then was thrown out of the airplane.

So the F9F-6 had that feature built in. You could arm your seat with the canopy still on by an emergency arming device next to your head. So in this negative "gee" situation that I was in, I tried everything to get the airplane out this unusual attitude, but it just kept going under. So with all those negative "gees," I was pulled almost up into the canopy, and the handle for blowing the canopy off was down on my left side. Being pulled up so high by these negative "gees," I couldn't push this handle down far enough to arm the seat and blow the canopy off. So there was only one thing to do—go *through* the canopy. It had never been done before, but somebody had to do it first, I guess. So I armed the seat and blew myself, seat and all, right through the plexiglas. Doing it that way probably saved my life because the plane was already through the speed of sound and if I'd have blown the canopy, the slipstream would have just whipped me to death. We had a case of an F-86 where a pilot had bailed out above the speed of sound, and he didn't live through it. They recovered his body, and his face was just torn up from the wind. But in my case, going through the canopy, I had started slowing down immediately as soon as I had left the airplane. I was slowing down, and I didn't have to sit there in the seat with a canopy gone and this Mach 1.0 slipstream coming through the cockpit, although when I hit the slipstream, it tore off my oxygen mask.

Here I was up at 32,000 feet with no oxygen, wanting to freefall because that was the only way to get down where I could breathe. I saw I was going to pass out, and I wasn't going to hit the ground passed out, so I deployed the parachute anyway. So I passed out, hanging in the chute, and of course we had been taught grunt breathing in training. There's oxygen at 30,000 feet; it's just the pressure's not there to force it into your lungs. So the basic idea is to suck in a big breath and force pressure on your lungs to try to force some oxygen into your bloodstream, which I started doing—and it worked. I would grunt breath and then force real hard and then be in the gray area, about to pass out again, and then I would clear up as that blood hit my heart with oxygen in it. So I did that grunt breathing down to 15,000 feet, where oxygen was plentiful enough for me to breathe normally. It took about twenty-two minutes from the time my chute deployed until I hit the ground, so I was up there for a while.

There were six of us all together, and the others saw me pitch out of the formation and down; so two of them started following me down. They never did see me bail out. They stayed with the airplane, because I bailed out upside down. They followed the plane all the way down until it hit the ground and exploded, and then they started climbing back up and saw me coming down in the chute. They came up and started circling me, and I was at 22,000 feet then.

I came on down in the chute, and hit in a cotton patch just outside of Pickens, Mississippi. A farmer was in the local area and saw me coming down. He had heard the claps from the airplane passing through the sound barrier up there, and looked up. There wasn't a thunderbumper around anywhere, and then all of a sudden he saw this plane screaming down, hit, and crash—luckily enough in a wooded area where it didn't bother anybody. So he came over to pick me up, and the planes were buzzing me back and forth to see if I was all right. I finally gave them a "roger" signal to shove off. I assumed I was okay; I was alive. They left and three of them went into Memphis; two of them went on into Barksdale in Shreveport. They had stayed up and passed out messages to Barksdale to let them know we had lost an airplane.

So the Highway Patrol came and picked me up in Pickens and took me to Jackson, Mississippi, where a navy plane came to pick me up and flew me into Memphis. They took me to the hospital and looked me over and said I was all right. I stayed in Memphis that night and got to Corpus the next day. I had a few bruised ribs and frostbitten ears, but other than that I was in great shape. I flew a show six days later, so I guess I was in much better shape than I should have been.

That I had the presence of mind to do all the things I had to do says something for navy training. Without the training you wouldn't make it. You've got to do the training, have it instilled in you, and when the time comes you follow that procedure and then away you go. At that time we did not have the modern things they have now—like chutes that deploy themselves and oxygen bottles and that type of thing for bailing out. Today you freefall and don't worry about it, because you have a barometer on your seat at a preset level, say, 5,000 feet, which pops the seatbelt and opens the chute for you; you don't have to do it yourself. During that time—1953—those things existed, but they hadn't started putting them in the aircraft yet. So now they have a chute that you can eject from the airplane at ground level—a Martin-Baker seat, a rocket-type seat that shoots you up high enough for the chute to deploy. Everything is done automatically; you just have to do the ejecting. When you decide to go, you go, and you depend on the automatic devices to take care of the rest of it.

After my bailout I was well known to all the younger bucks coming through because the taped version of the bailout became required listening for all students coming through flight training. I wouldn't be surprised if they're still playing the tape every so often. Of course the flight surgeons are always telling you what you did wrong, you see. "So here's the things he did wrong." But he's alive, so, "Here's the things he did right." There were a lot of things done wrong, no question about that—such as not having a bailout bottle in the parachute I had. We did have the bottles, but our chutes that we left at home for this trip had the bailout bottles in them, although in my case it wouldn't have mattered anyway because the slipstream tore off my oxygen mask when I went out. Also this was just a cross-country flight, so I didn't have my g-suit on. I just had my flight suit, which might have helped to save me, because under that I had my regular uniform on and I didn't freeze to death hanging up there at 30,000 feet, although my ears got frostbitten.

So there were lots of things that the flight surgeon could show that I did wrong, but some things he could show I did right. From the tape a story was written for *The Saturday Evening Post* and it was reprinted in about fourteen different countries, and picked up in books—my only bailout during my thirty-one years of flying.

===== 5 =====

Carriers for the Jet Age
Angled Decks, Steam Cats, and Mirrors

Admiral James S. Russell

James Sargent Russell was born in Tacoma, Washington, on 22 March 1903, son of Ambrose James and Loella Janet (Sargent) Russell. After graduation from high school and a tour as a seaman in the Merchant Marine, he entered the U.S. Naval Academy in 1922, was graduated and commissioned an ensign in 1926, and was designated a naval aviator in 1929. There followed normal tours of aviation duty both on board ship and ashore, including two years in the Bureau of Aeronautics, during which he was closely associated with the design of the Essex-*class carrier.*

Following distinguished service in combat in the Pacific during World War II, Admiral Russell served in many operational and staff assignments of great responsibility, including commanding officer of the USS Coral Sea *(CVA-43), commander of carrier divisions in the Western Pacific, chief of the Bureau of Aeronautics, deputy commander in chief of the Atlantic Fleet, and vice chief of naval operations, as he advanced in rank to admiral on 21 July 1958. Admiral Russell became commander in chief, Allied Forces, Southern Europe, on 2 January 1962 and served as such until his retirement, effective 1 April 1965.*

During his illustrious career, Admiral Russell was awarded the Distinguished Service Medal with one Gold Star, the Legion of Merit with two

Adm. James S. Russell, commander in chief, Allied Forces, Southern Europe, 1962–65. Former chief of the Bureau of Aeronautics, 1955–57.

Gold Stars, the Distinguished Flying Cross, the Air Medal, the Navy Unit Commendation Ribbon, numerous campaign and service medals, and several foreign decorations.

In 1952 I was detached from command of the USS *Coral Sea* and came back to the Air Warfare Division in the Office of the Chief of Naval Operations (OpNav) under OP-05, the head of aviation in the U.S. Navy. This was essentially my old division of military requirements in the Bureau of Aeronautics. I was there for a regular tour of two years, and while there, I was selected to flag rank in the spring of 1952.

There were some new developments with which I was involved, both in carriers and in aircraft. In the carriers, there were tremendous advances: the angled deck came in, along with the steam catapult, and the mirror landing system, which provided an optical glide path. The fact that the deck was angled allowed one to do away with the barriers, which used to be the cause of an airplane's demise, because if you overshot the arresting gear, you crashed into the barrier.

ADMIRAL JAMES S. RUSSELL

The British were responsible for this new development. A very good friend of mine, Rear Adm. Dennis Campbell, RN, was a leader in this. It seems he told his secretary to put her lipstick on the left side, the port side, of a table, and to put a pencil beyond it. Then he told her to make a circuit around the room, and when she was lined up with the major axis of the table, keep the top of her lipstick container in line with the lead of the pencil lying on the table. This the girl did and she wound up with her chin on the edge of the table. Whereupon Campbell said, "See, this is the mirror landing sight. It brings you down very accurately to the deck."

Well, in fact it did, and then we got Fresnel lenses as a development of the mirror landing sight. Primarily from the pilot's viewpoint, you see a bar of green lights with a space at the center, and then you see a yellow light in the center. The yellow light is called the "meatball." Aboard ship, because of the pitch of the ship, you have to have it gyro-stabilized. But this yellow light, if you're high, is above your broken line of green lights; and if you're low, it's below. So when you come aboard, you keep this yellow light between the ends of the two green bars, and this means you are on a glide slope, which is usually about three degrees to the horizontal, and you don't even flare the airplane—you fly into the deck.

Carrier-based airplanes have a stronger landing gear than land-based planes, and they can stand the impact of the three-degree descent when they touch the deck. This system was particularly needed for jets, because with flaps, wheels, and hook down in a jet, you are fairly aerodynamically dirty, and you're drawing about 80 percent of your jet thrust just to keep flying at a steady speed, about 5 knots above stalling speed. Coming down on the optical glide path, the moment you touch the deck—even though you catch a wire with your arresting hook—you immediately apply full throttle. This gets your turbine up to full speed so that if you don't have a wire, that full throttle will give you thrust to take off and come around and try again. That's called a "bolter," if you miss the arresting gear. If you catch a wire, there is no problem. In an arrested landing, so far as the arresting gear is concerned, it matters little whether your throttle is wide open or closed; the dominant force is that of slowing the mass (weight) of the airplane down.

It sounds simple, but when the angled deck came, I wasn't completely sold on it, until I realized its great virtue in handling jets, with which we'd had all sorts of trouble. A jet was so structurally smooth that when it hit the barrier, sometimes it went right on through and crashed into the airplanes parked up ahead. So here was a way of solving that problem.

Now, the catapult was another matter. We had gone from compressed air to gunpowder as a launching source of energy. But in carriers, if you had a powder charge to expend every time you launched an airplane, you'd soon sink the ship carrying powder charges. So something had to be done, and it was done in this fashion. We had an accumulator with compressed air, which was used only as a spring on top of hydraulic fluid. The hydraulic fluid under pressure was admitted through a quick-opening valve behind a piston, which, through a system of wires, towed the airplane into the air. The piston was stopped by hydraulic dash pots with plungers.

The business side of the catapult was a shuttle in a slot in the deck. Power was applied by the hydraulic fluid pushing a plunger out, which pulled the cables that drew the shuttle forward to launch the airplane. This meant that instead of wasting energy by exhausting the launching pressure to the atmosphere at the end of every catapult stroke—as you would with a powder charge or with compressed air—you conserved the elastic medium, because when you reached the end of the stroke, you could cut off the pressure by closing the launching valve. Then retracting the piston, you just drained an incompressible fluid into a sump from which it could be pumped back up into the accumulator against the pressure. So you saved all that energy and thus made catapulting a very common, practical thing for a ship.

So we went along with the hydraulic catapult from the days of the *Yorktown* (CV-5). But with the ever-increasing size of the airplane, the ever-increasing velocity at which you had to launch it in order to make it fly, we found we needed to do something to get away from all the wasted energy in starting and stopping.

Again the British came to the rescue. They developed the steam catapult in which a piston bearing the shuttle was driven down the deck by steam. Thus, the launching force was more directly applied to the airplane. They did it by having a closed cylinder behind the piston and an open cylinder ahead of it. There were two steel ribbons that lay on the bottom of the tube, so you had an open cylinder forward of the shuttle; but the shuttle picked these two steel ribbons up and sealed them, so that you had a closed cylinder behind the piston.

Furthermore, you controlled the acceleration by injecting steam through a series of ports as the piston moved along the deck. The object, of course, was to give a steady acceleration to the airplane, to avoid peaks that would give impact loads to the airplane structure. Here was a piston, directly hooked to the airplane, so there was practically no inertia to stop at the end of the catapult stroke, except the mass of the piston itself, and the shuttle, of course.

Well, those were three tremendous improvements—angled deck, steam catapult, and the optical glide path. While I was on the desk for military requirements and new developments, there were some new planes that were being designed—the Vought F8U, Douglas A4D, and the McDonnell F4H. But it wasn't actually until 1956, when I was at the Bureau of Aeronautics, that we got the Collier Trophy for the first ship-based airplane that flew faster than 1,000 miles an hour—the F8U. It took about four and one-half years from its design inception to first service use.

After my two-year tour in the Air Warfare Division, I returned to sea in May 1954 as commander, Carrier Divisions 17 and 5. In March 1955 I reported as chief, Bureau of Aeronautics, in time to defend a budget before the Congress, a budget that I had not seen before—a rather difficult job. The budget was the navy procurement budget for airplanes for fiscal year 1956, as I remember—probably on the order of three billion dollars. It was difficult to shepherd this through the Congress. We, of course, had many hearings on the regular budget, but we were also beset with investigations. We had a number of aircraft designed to take what was to be known as the J40 engine. The Westinghouse Company had done very well with an axial-flow engine called the J34, and the J40 was to be of similar design but greatly increased in power. It was a very compact engine in its design. Westinghouse fell down badly on the job; the engine came out greatly reduced in power. It was the only engine that would fit into the F3H-1, a McDonnell fighter.

When we started losing airplanes because of the engine, I, as chief of the Bureau of Aeronautics, gave the order to ground the aircraft. This was done, and my staff informed me that there were twenty-six F3H "gliders" at the McDonnell plant in St. Louis, and what did I choose to do with them. I said, "Well, we have a mechanics school at Memphis. Let's put these airplanes on a lighter and take them downriver to Memphis and let the mechanics assemble and disassemble them to find out how a modern airplane is built."

We were towing these aircraft through the streets of St. Louis down to the waterfront to put them on a lighter in the dark of night when we were discovered by the press. Headlines broke out across the country about the navy's abysmal lack of appreciation for good business, that we had invested in an airplane that couldn't fly or didn't have an engine to make it fly. It was immediately made a subject of congressional investigation. The airplanes were, in fact, loaded on the lighter and taken down to Memphis and served a purpose, certainly not the intended purpose. But to add insult to injury, the reporter who discovered our towing act won the Pulitzer Prize for his reporting for the year.

Thinking of how to defend this fiasco before a Subcommittee of the House Armed Services Committee of the Congress, I decided that the best thing to do was to confess our error and quite honestly face the problem that one has when designing a high-performance fighter. I explained to the gentlemen on the committee that a second-rate fighter was no fighter at all, and a fighter really had to embody the very latest in technology. In doing that, one had to run certain risks, and occasionally, these risks turned out to be very real. In this case, the navy made a mistake in putting their faith in the J40 engine.

The chairman couldn't adapt himself to the fact that I was admitting a mistake. He rose in his chair and glared down at me like the bull in the pit; they sit in an elevated circular arrangement, and you're down below in the pit. He rose from his chair, pointed his finger at me, and said, "Admiral, do you mean to say that you admit that you made a mistake?"

I said, "Yes, sir. We made a mistake," and that sort of took the wind out of the sails of the investigation. We produced all the figures they asked for, how much money had been spent in the development of the engine.

Other manufacturers such as Douglas, with the F4D, had left a big enough hole in the fuselage to put an alternative engine in—the Pratt & Whitney J57—but McDonnell had economized in space to the point where no other engine would fit this particular fuselage. McDonnell did come out with an F3H-2, a succeeding model, and they were flown reasonably successfully, but the only engine that would go into that was the Allison J71. And we had some trouble with the Allison turbojet flaming out if it was flown through a driving rainstorm.

So, not only did I have to defend a budget, which I had no hand in preparing, but I had two rather serious investigations to testify before. One was the J40 engine itself, and the other was the F3H-1, an airplane that would take no other engine than the J40.

Of course, time went along and I had other congressional investigations. I remember very well when Senator Stuart Symington decided that there was an airplane gap—that the U.S. Air Force was not receiving the attention it should with better airplanes, and the United States was falling behind. He'd very carefully planned this, but President Eisenhower insisted that if the good senator was going to investigate airplanes, Senator Symington must, of course, include the naval air force in the investigation. So, we came to that particular investigation with two strikes against us, because we'd been thrust upon the committee unwontedly. I'm sure that we didn't care to appear, but because of the rather one-sidedness of Senator Symington's investigation, the president had told him that he must include the navy in his investigation.

One very interesting thing came up while I was there: the matter of guided missiles and ballistic missiles. The combination of a nuclear-powered submarine and a ballistic missile was a very intriguing thing. The Bureau of Ordnance tried to push a study of ballistic missiles, but they had an air breather that was their favorite at the moment—a supersonic air-breather missile. I think they were really fearful that the project might be damaged by a truly ballistic missile. It so often happens—particularly if the Congress gets wind of something that might be better than what you have aboard and what you're asking for—they will postpone that in the name of economy to get something better. It's the old story of the dog and the bone. You know, the dog saw the reflection of his bone in the water, went for it, and lost the bone.

We in the Bureau of Aeronautics were very intrigued with the idea of putting ballistic missiles in submarines, but we had a couple of handicaps. One was that the Bureau of Ordnance was very backward about ballistic missiles. We had difficulty, because Adm. Arleigh Burke was now the CNO, and he tended to support his old bureau, and beyond that, Mr. Charles Wilson, secretary of defense, said, "Yes, we could have a ballistic missile on a submarine, but we'd have to use the army's Jupiter rocket." Well, the Jupiter rocket was a liquid-fueled rocket, and you can imagine nothing nastier than having liquid fuel loose in a submarine, and it was also too large. But I got sort of fed up with this, and I made contracts with three aircraft companies to study the problem of putting ballistic missiles in submarines. That sort of forced the hand and really brought attention to bear on the subject.

This became a practicality when the solid-fuel ballistic missile turned out as well as it did. Polaris became a project, and then it was a matter of who would take this project. The Bureau of Aeronautics had been pushing the project, and yet the rocket, one might say, should belong to the Bureau of Ordnance, since it concerned a piece of ordnance rather than an airplane. This case came before Mr. Thomas Gates, who was secretary of the navy. Mr. Gates joined in a conference among us to pass judgment on whether it would be put in Ordnance, whether it would be put in Aeronautics, or where it would be put. He led off with, "I believe in giving projects to the people who work at them."

I thought to myself, "Well, he's going to give it to the Bureau of Aeronautics."

Then he said, "One might think that this is in the province of the Bureau of Ordnance. But a curse on both your houses! I'll have it directly under me. However, because Aeronautics has pushed it, we'll put an aviator in charge of it. But the Polaris project will report directly to me."

That was the genesis of the Special Project—the first Special Project in the navy. Red Raborn, an aviator, was taken from the Guided Missile Division of OpNav and put in charge of it. A very astute ordnance engineer by the name of Levering Smith was his number two, and the master technician on the job. They set up special offices and had carte blanche to take any Aeronautics and Ordnance people whom they desired for the project. This was used with great effect. A lot of our very good people were shifted over to the Polaris project. The Program Evaluation and Review Technique (PERT) system was developed by a civilian, Gordon Pehrson, so that the entire program could be managed efficiently. Polaris was off and running, and it became a great success.

Other than personnel, the Bureau of Aeronautics contributed the three studies that we had made. Ordnance was given the housekeeping chores. Special Project number one set up shop in a wing in the munitions building. Aeronautics, of course, was pleased that this project could be brought to fruition, as indeed it was adroitly and competently done by the team assigned to it.

Secretary Wilson was convinced, finally, when the ballistic missile with solid fuel became a reality. Of course, the solid-fuel Polaris rocket put the U.S. Air Force on notice that they would have to do better than their big liquid-fueled rockets, and they too, then, went into a program known as the Minuteman program. But it was really kicked off by the fact that Polaris turned out to be such a success.

One of the interesting projects of that time was the P6M flying boat, the Seamaster, which Martin was developing. It was developed when I was chief of the Bureau of Aeronautics. The tips of the wings drooped downward and acted as wing-tip floats. The hull itself was very sleek looking. Truly, it was a beautiful airplane. It had a bomb bay that rotated, so that when the aircraft was performing as a flying boat, there was a continuous surface underneath the airplane sealed by inflated gaskets so that it was water tight. When it was airborne, the bomb bay, which was in cylindrical form, just turned around, so an open side of the cylinder was down, and you could drop your bombs through the bottom of the seaplane.

Much to our sorrow, an aircraft was lost under rather peculiar circumstances. It was in the higher part of the speed envelope approaching the speed of sound, and the airplane "tucked"; it had pitched down with such violence that the wings, under negative "gee," or downward forces, had come off. The airplane crashed and everyone in it was killed. Of course, an investigation was undertaken to find out what the trouble might be. A second Seamaster was being tested when another failure in flight took place. This time the aircraft pitched up and the crew safely ejected. After the people ejected in the pitchup, the aircraft slowed, seemed to recover, and almost flew itself down to the ground.

It was discovered that there was a rather simple arithmetical error in scal-

Revolutionary and pleasing to the eye, the ill-fated Martin P6M Seamaster was beset with difficulties during its short-lived development program. (Courtesy National Air and Space Museum)

ing up the control forces from wind tunnel data. What was happening was in the hydraulic control. If the nose of the airplane started up, the pilot would push his yoke forward, and what he was doing was opening a hydraulic valve intending to put pressure on the control surface to make it counteract the rising of the nose. But the force on the control surface overcame the pressure of the hydraulic fluid and actually forced it back through the system. So all you were doing in trying to control the airplane's attitude was opening a valve that allowed the control surface at the tail to go to an extreme position under aerodynamic loads in a direction opposite to that desired. Thus a violent tuck, or pitchup, resulted. This unfortunate condition was met by putting more power in the elevator control system.

There were several other serious deficiencies in the aircraft. Because the wing was so close to the water, wing flaps could not be used. Without wing flaps landing speed was uncomfortably high, particularly when operating in a rough sea. Taxiing on the water was a problem because thrust reversers could not be used. Thrust reversers would have kicked up enough spray and water to drown out the jet engines. Without thrust reversers, all waterborne maneuvering had to be in a forward direction.

It should be noted that the Seamaster had a sizable water rudder. It served not only as a water rudder but could split and fan out to be used as a water brake. For beaching, the Martin Company developed a wheeled cradle into which the Seamaster could be taxied. With a hydraulic hookup to the aircraft, the cradle could be steered and braked from the aircraft. Thus, aircraft and cradle could taxi up a seaplane ramp and into a spot on the parking area.

Beset with many difficulties, and after limited production had begun, the P6M program was finally terminated. Secretary of the Navy Gates was one of the last to consent to the cancellation of the project.

One of the more interesting airplanes was the A3J, a reconnaissance plane built by North American in Columbus, Ohio. A beautiful-looking thing, it was not the best carrier plane, but it developed into the RA-5C, with a tremendous photoreconnaissance capability. It was a real step forward in reconnaissance. We had reconnaissance fighters: we'd adapt fighters to reconnaissance and call them RFs, but the RA-5C was the first real advance in reconnaissance since World War II.

In August 1957 various people got interested enough in my career as chief of the Bureau of Aeronautics to suggest that I should go to sea and get groomed for higher things, and I was, in fact, assigned as deputy commander in chief, U.S. Atlantic Fleet.

6

1958

The Transition Year

Vice Admiral Robert B. Pirie

Robert Burns Pirie was born in Wymore, Nebraska, on 18 April 1905, son of Charles Bruce and Thelma (Harms) Pirie. He attended Wymore High School, was appointed to the U.S. Naval Academy, and was graduated and commissioned ensign on 3 June 1926. He subsequently attended flight training at Pensacola and was designated a naval aviator on 25 June 1929. Subsequent service included extensive command and staff duties ashore and afloat. He advanced in rank to vice admiral, to date from 3 July 1957. He served as deputy chief of naval operations (air) from May 1958 until his retirement, effective 1 November 1962.

During his distinguished career, Admiral Pirie was commandant of midshipmen at the U.S. Naval Academy, deputy commander in chief, Naval Forces, Eastern Atlantic and Mediterranean, chief of staff and aide to the commander in chief, U.S. Atlantic Fleet, commander, Carrier Division 6, and commander, Second Fleet.

Admiral Pirie's decorations include the Distinguished Service Medal, the Silver Star, the Legion of Merit with Combat "V" and Gold Star, the Bronze Star with Combat "V," and four Presidential Unit Citations.

I reported to the Office of the Chief of Naval Operations in May 1958 to assume my duties as deputy chief of naval operations (air). When I reported to

*Vice Adm. Robert B. Pirie,
deputy chief of naval
operations (air), 1958–62.*

Adm. Arleigh Burke, the chief of naval operations, he told me that there were two very important things he wanted me to accomplish during my tenure as DCNO—first, to manage the business of getting a carrier authorized; second, to know more about the budget than any of the other deputies or anyone in the CNO's office. As a result of these two things that he told me he wanted me to pay particular attention to, during my period as the DCNO we had two aircraft carriers authorized to be built and I did make a very thorough study of the budget, and I believe I did know as much or more about the budget than any of my contemporaries there at the time.

The aircraft carrier always was under heavy criticism from opponents who thought that it is very vulnerable to attack and that to place much of your money in a vulnerable weapon system was not sound. We aviators didn't agree with that. Nevertheless there were great opponents to the aircraft carrier from the very beginning, and after World War II they were even stronger in their opposition because of new weapons systems that were coming into being and they thought that we couldn't defend the aircraft carrier.

A lot of this opposition was in the Congress itself, so that it was necessary to work hard in the Congress to be sure that you had a majority who were for the appropriation to build an additional carrier.

The year 1958, in terms of naval affairs, has often been called a transition year, when the navy was turning from guns to missiles, going into the realm of nuclear power, and it was going from subsonic to supersonic speeds in the air. New and fascinating systems, including the Polaris missile, were under full development. All of these had a great effect on our budget considerations and on what types of aircraft, missiles, and ships we were going to buy. The nuclear submarine program, which was in its infancy at that time with the commissioning of the *Nautilus* (SSN-571) imminent, also had a great bearing on weapons systems development.

A great many new and interesting aircraft developments were taking place at the time. Shortly after I took over, we had to make a decision as to what fighter we were going to buy, whether it was going to be the F4H— now known as the F-4—or whether we were going to buy the F8U-3, a product of the Chance Vought Aircraft Corporation, now LTV. The F8U-3 was a single-engine fighter that in performance and handling characteristics was a little bit better than the F4H at the time, according to the preliminary evaluations of the test pilots, but one of the basic reasons for our taking the F4H was that it had two engines. My philosophy was that two engines in every combatant aircraft were better than one, because you always had an opportunity of getting back with one engine. When an engine fails in a single-engine supersonic jet you are almost called on, if you can't get the engine started, to eject and the plane is lost. The cost of these modern weapons systems and supersonic aircraft had become so great that this was a primary consideration in our final decision. We chose the F-4, and it's been a very successful aircraft, not only in our own navy but throughout the world.

So that decision was made shortly after I took over, and the F4H (F-4) then went into production. We had a number of other aircraft coming along—the A2F (A-6), which has been a very successful attack airplane; newer and better versions of the A4D (A-4); the A3J (RA-5) reconnaissance aircraft; the A2F-1H (EA-6A) electronic countermeasures aircraft; the W2F-1 (E-2) airborne early warning aircraft; several versions of helicopters. All of these were very expensive projects and we had to work hard to save them.

The aircraft nuclear propulsion program was one of the most interesting and controversial when I took over in 1958. We had an office in my establishment looking to the development of the aircraft nuclear propulsion plant to go in a large seaplane. We had two companies developing aircraft nuclear propulsion plants to produce the energy to run an engine and drive a pro-

For two decades the McDonnell Douglas F-4 Phantom II was the world's premier fighter-bomber, flown by the U.S. Navy, Marine Corps, and Air Force, and with the air arms of ten other free-world nations.

peller or a jet to power the aircraft. The aircraft we had in mind at the time—without going through a completely new development of a large aircraft—were three Princess flying boats the British had in mothballs. They were in good condition and large enough to be able to handle the weight we would require in an aircraft nuclear propulsion system. The reason the weight had to be so great was to shield the power plant from the crew so that it wouldn't affect them. That required a sizable aircraft.

I appeared before Congress, particularly the Joint Atomic Energy Committee, to try to get authorization to continue the development of these aircraft nuclear propulsion systems with the possibility of leasing or buying these Princess boats from the British. This program was not supported by Admiral Rickover, and as a result of his opposition, in my opinion, the program died. It was not invented by him. I've told him that to his face, so I'm not saying anything out of school. In my opinion, had they let us continue this development at what I considered a reasonable cost at the time, we could have had a nuclear airplane flying ten or twelve years ago. However, the program was discontinued and has long since been abandoned. The air force had a similar program, which also died aborning, to power a large land-based bomber with nuclear propulsion.

I'd like to go back a little in history and talk about the development of naval aircraft that we've had in the past two decades, how the development came about, and the people who were the most responsible, in my opinion, for this development, because I believe firmly that the U.S. Navy has the best military aircraft for the purpose intended in the world. To go back to the aircraft development after World War II, the four chiefs of the Bureau of Aeronautics who I believe were significant in the development of aircraft and engines were Mel Pride, who had a great deal of World War II experience; Apollo Soucek of record-breaking fame in his early days and who had quite a World War II record; Jim Russell, who I think is the principal and most significant individual in the development of aircraft for the U.S. Navy; and his successor, Bob Dixon. Jim Russell and Bob Dixon spent a great deal of time in the Bureau of Aeronautics during their careers and I believe they're the principal contributors particularly to the carrier aircraft development that has been so successful.

I might name two or three individuals who passed through the flight test section of the navy and had a great deal to do with the quality of aircraft the navy had. First, Fred Trapnell, who I believe personally is the best pilot I've ever been associated with, and probably the best test pilot; we were both in the test section back in 1931–32. His successor was William V. Davis and they together contributed a tremendous amount to the development of naval aircraft through the test and evaluation program. They set up those programs at Patuxent River (Maryland) prewar, during the war, and subsequent to the war. Tom Connolly, who was a protégé of Trapnell, was the first head of the flight test school at Patuxent. They all had a significant part in the development.

Another individual who had a great part in the engine development was Bill Schoech. Bill had his postgraduate training at Cal Tech in engines and he, I think, probably as much as anyone is responsible for the fine engine development, particularly through the Pratt & Whitney Company and United Aircraft.

The development of aircraft with turbojet engines, which permitted us to have a significant increase in aircraft performance, took place after World War II under the guidance of these individuals I have mentioned. Most of the fighter aircraft were developed by three companies—Grumman, McDonnell Douglas (McDonnell at that time) in St. Louis, and North American under Dutch Kindelberger. The fighter aircraft that was probably the most successful up to the 1970s is the McDonnell F-4. The Grumman F-14, I believe, was the finest fighter airplane of the 1970s; it is by far the most sophisticated in every respect in its ability to deliver weapons in any mode. By this I mean it could deliver standoff missiles of great capability, it could dogfight with cannon or with short-range missiles. It was very maneuverable, and could come aboard a carrier 25 knots slower than the F-4.

We've tried very hard to keep aircraft approach speeds within reason because of the material aspects of getting a plane on board a carrier. Landings at very high speed require great strengthening of the aircraft, its landing gear, the tailhook assembly for arresting, and the ship's arresting gear cables and engines. The lower we can get the speed, the less the requirements in strength and weight for arrested landings. It's far easier for the pilot to get aboard under all conditions with the speed down at some reasonable level. Advances that have taken place over the past ten years permit us to get lower landing speeds.

Still speaking of the 1970s, in the attack plane field we had two great airplanes, the A-6, which was developed specifically for accurate weapons delivery by the Grumman Corporation, and the A-7, which was the Vought aircraft development that was a very significant airplane in our suit then. In the antisubmarine plane field Lockheed was building the new S-3 twin-engine jet for carrier operations, and I think it was probably the finest airplane in its field. The patrol plane we were using, the P-3, was a splendid long-range antisubmarine warfare craft and I think the system that goes with it is still the finest antisubmarine weapon in the world in the 1990s.

The development of the electronics systems for each of these aircraft kept pace with the actual airframe and engine development, and in antisubmarine warfare the systems used for searching out and finding the submarine as well as holding contact with it until you can destroy it are extremely effective, as are the destructive systems. I believe that the aircraft system is a better system than any surface or subsurface system in existence today for the same purpose. I think the S-3 has an equally effective system in a smaller airplane that can fly off the carriers.

In the field of fighter aircraft we have new aiming systems developed around laser beams that are most effective and can pinpoint another aircraft and a ground target, and they are being used in our attack aircraft for pinpoint accuracy. All of these developments came from a very responsible early education system. Most of the officers that I have mentioned had postgraduate training in aeronautical engineering and had the real background knowledge and foundation to become experts in the field.

The time process involved from the beginning when a weapon is conceived until it goes through the various phases of development and finally reaches the fleet is on the order of five years. That's a ballpark figure. It may be a little longer than that from the concept, but from the time they decide on the development until they actually get the things flying in the fleet it's roughly about five years.

The expense of our aircraft weapons systems has come in for a great deal of criticism in the past few years, and I think with some justification. In our effort to give the pilot and the crew of each of our combatant aircraft the most significant and best systems we can give them to do their jobs, we get a great many research, development, and engineering personnel working on these systems, and by the time we get the systems into operation they are very expensive. The only way you can cut the cost is to take away from the aircraft a significant system for fighting, which I don't think is justified.

The devastating destructive power of these modern weapon systems showed up in the 1973 Israeli-Arab war in which large numbers of aircraft, tanks, and missiles were destroyed in the shortest period of time ever recorded in history. I think that kind of accuracy of both the offensive and defensive systems is going to rule the next conflict we see.

The training system we use for pilot and aircrew training, I think, is most significant in naval aviation and has been from the beginning. From the real inception of naval aviation, starting in World War I, and then in the interim period between World War I and World War II, when the carrier came into being and became a significant weapon system, we had to develop techniques for carrier operations, and the training kept pace with it. I think that as proof of our system we had a significant number of the astronauts chosen from among naval aviators.

In training flight and ground crews we found in the early stages of naval aviation that it was most important to devote a lot of time and effort to training programs vis-à-vis just training under operational conditions, and the technical training programs that we developed at Memphis have really paid off. I can remember when I was DCNO trying to convince the chief of naval personnel that they should do the same for the surface navy, particularly in the field of firing surface-to-air missiles, and that they should adopt our methods. He always said our methods cost two or three times as much as their methods of training and they couldn't afford it. I said, "Yes, but our people are trained." It's significant that in both aircraft and submarine training it's a matter of survival. If you have poor crews you're going to lose lives and airplanes, and the same thing happens on a submarine. The surface people never quite got around to that because they weren't faced with survival just from operations in peace or war. They did most of their training on the job, instead of having significant training at a technical training center.

I also think our postgraduate training is excellent. It's done professionally and at some considerable expense in the sense of taking officers out of operational assignments for two or three years in order to do it. But it's significant that we do, I believe, have the world's finest personnel, aircraft, and ancillary equipment required for these aircraft for both offensive and defensive operations.

There were a good many arguments and battles over numbers of pilots and flight crews to be trained, and I had an opportunity to work with the air force in arguing with Defense Department officials about this. I remember distinctly that in the early part of the Kennedy administration, Robert McNamara and his crew thought we should reduce the pilot training rate, and I worked very closely with Gen. Emmett "Rosey" O'Donnell, who was then the deputy of the air force for personnel. We argued against really significant cuts with the thought that it is very difficult to regenerate and start up a training program, and it's far more expensive to get into a conflict and start up a training program than it is to have a continuing program of significant numbers.

They made the cuts, arbitrary cuts. They cut the air force far worse than they did the navy and marines, but they made some arbitrary cuts and cut our training rate down, and, lo and behold, a year or two later along came the Vietnam conflict and it all had to be turned up again and started, and the costs were significantly greater than had they continued. The thing that we fought for was to train larger numbers than you need and hold the ones that leave the service as reserves in readiness to be called back. Say, if your training rate is three hundred a month and you only need one hundred fifty, you put the one hundred fifty that you trained every year on the shelf, and they're your reserves. That's a cheaper system than it is to go back to one hundred fifty with a smaller training establishment and jack it up to a higher number when you get into a conflict. This is a hard concept to sell at the moment.

Also you've got to get the pilots and the ground crews and the aircraft that are necessary to train these people, and to try to accelerate takes a significant length of time. So, if you get into a conflict it's two years before you get the first product of an accelerated program. Suppose you're at a pilot training rate of a hundred and fifty a year and you have to go to five hundred; well, you stay at that one hundred fifty that's always coming off the end of the line for the first two years. It's two years before you get anything more, if you're lucky—if you can get the aircraft and the necessary pilot- and crew-training personnel. This was always a big argument and O'Donnell and I got thrown out of two or three offices more than once, arguing this case. But the proof was in the pudding when the Vietnam War came along.

It's very important in the operation of any of these sophisticated systems in aircraft today that we have adequate spares. The provisioning and finding out what the use of each part of the plane's equipment amounts to so that you can buy intelligently is quite a job. It's necessary to deploy what we call "provisioning teams" with the aircraft during its development, much earlier than we did three or four years ago. They have to find out what the usage rate is, so that we don't buy too many of a part that we don't need, and we do buy a significant number of those that have high usage. In these expensive systems today this becomes one of the most important aspects of keeping your systems operating and doing it at a reasonable cost.

I don't like to be known as the undertaker who drove the last nail in the coffin of lighter-than-air. However, it fell to my lot, and not too pleasantly, to evaluate and decide what to do about lighter-than-air. It was a costly operation and, for value received, we had to make a careful analysis of what lighter-than-air could contribute vis-à-vis heavier-than-air to our national defense. We did make such studies and found that it should be eliminated as a system because it wasn't effective as compared with heavier-than-air, so the program was cut significantly and finally put to bed during my term as DCNO.

I don't think it's an irrevocable decision if someone can figure out a use for them, in the scheme of things, perhaps in antisubmarine warfare. The actual record of blimps in antisubmarine warfare during World War II was that they didn't sink one German submarine and that one lighter-than-air craft was destroyed by a German submarine. That was their total record. They flew a lot of hours and if you want to postulate that they did keep the German submarines underwater instead of coming to the surface, they might have been valuable in the coastal areas. They're of no value in bad weather.

I operated them once, from the carrier *Sicily* (CVE-118) when I had command. We experimented with making landings and flying the blimps from the ship off the East Coast, and we had quite an extensive exercise down off Vieques near Puerto Rico for five or six days, operating them continuously away from any other base. We fueled them and provisioned them down on the decks in calm weather. I think there is a limit as to what kind of weather you can operate them in. One of the primary reasons for recommending their discontinuation was the accidents and significant losses due to the weather. I think that's what destroyed the original *Akron* and *Macon*. So it was very difficult to see what their value was compared with the large aircraft for the same amount of area and effectiveness.

Shortly after Mr. McNamara came in as secretary of defense and appointed Alain Enthoven as the head of a new section within the controller's office to do systems analysis, it became evident that they were going to what I vulgarly like to call "mess around in my playground." They started wanting detailed analyses of why we picked certain aircraft, and tried to tell us how to design and develop the next generation of aircraft, so we had many real confrontations with them on aircraft systems. My particular dislike for them and for their approach to the problem was that none of them had any significant experience. Most of them were thirty-year-old Ph.D.s who'd never been in the military, never had any military experience, never had any experience with aircraft or aircraft engines, or anything to do with them. Just because they'd been hired by the Rand Corporation or some other "think tank" before they got to the systems analysis group didn't qualify them to try to tell us how to do our business. I used to particularly express my views in pretty strong terms that experience and a lifetime in this business should be given some consideration in the design of future naval weapons systems. We succeeded partially in getting our viewpoint across. It's significant that now each service has its own think tank and the business of evaluating weapons systems is now done within each service before the matter gets to the Defense Department.

I think that trying to second-guess the next major conflict in the world is something beyond the capacity of humans and, as Thucydides said, you want to reckon that the enemy's capability and intelligence are much the same as your own and you want to be prepared for any eventuality. The strong, courageous men who are willing to keep their convictions and do their job are more important than the storage of useless knowledge.

7

The TFX

One Plane for All

Vice Admiral Thomas F. Connolly

Thomas Francis Connolly was born on 24 October 1909 in St. Paul, Minnesota, son of Thomas Ignatius and Leona (Gillespie) Connolly. He was appointed to the U.S. Naval Academy in 1929 from California, and was graduated and commissioned ensign in 1933. After two years of sea duty, he commenced flight training at Pensacola, Florida, and was designated a naval aviator on 22 July 1936. Subsequent service included extensive squadron, command, and staff duties ashore and afloat, and he advanced in rank to vice admiral. Vice Admiral Connolly was deputy chief of naval operations (air) from 1 November 1966 until his retirement, effective 19 September 1971.

Following operational tours in observation and patrol squadrons in the 1930s, Admiral Connolly studied aeronautical engineering at the Naval Postgraduate School and the Massachusetts Institute of Technology, receiving a Master of Science degree in 1942. During World War II, he commanded Patrol Squadron 13 in the Pacific and was assistant director of the Flight Test Division at Patuxent River, Maryland. Subsequent command duties included director of the Test Pilot Training Division, commanding officer of Heavy Attack Squadron 6, USS Corregidor *(CVE-58) and attack carrier* Hornet *(CVA-12), and commander, Carrier Division 7. Duties with the Department of the Navy included director, Strike Warfare Division, assistant*

Vice Adm. Thomas F. Connolly, deputy chief of naval operations (air), 1966–71.

chief of naval operations for Fleet Operations and Readiness, and special assistant to the deputy chief of naval operations for Fleet Operations and Readiness. On 31 October 1965 Admiral Connolly assumed the post of commander, Naval Air Force, U.S. Pacific Fleet, and on 1 November 1966 he became deputy chief of naval operations (air).

Admiral Connolly's decorations include the Legion of Merit with Gold Star, the Distinguished Flying Cross with two Gold Stars, and the Air Medal with two Gold Stars.

On 1 November 1966 I reported as the deputy chief of naval operations for air warfare (OP-05) in the Navy Department. I knew that the number one problem that I was going to have to take on was the F-111, Secretary of Defense McNamara's TFX. The air force was going to use the F-111A as a low-altitude interdiction (attack) airplane, and the navy was to use the F-111B for fleet defense as an interceptor of bombers and missiles. Before I had left California to come to Washington, I had gone to Hughes Aircraft Corporation in Los Angeles. They had heard that I was going to OP-05 and at that time they had done relatively little business with the navy; they were mostly an air force contractor. The man who briefed me on the missile system for the F-111B was Dr. Malcolm Currie, who eventually became director for defense research and engineering in the Defense Department. The whole Hughes hierarchy was in that briefing room. At the conclusion of the briefing I remember saying, "Why are we talking about shooting bombers down, gentlemen? They're not the problem. The problem is going to be missiles—air-to-surface, surface-to-surface missiles. This system's great, but can you make it shoot down missiles?"

I can see their faces to this day and I can tell you their names; they were simply startled. I said, "This is a controversial program and it swallows a lot of money. It's a big switch from the way navy carrier operations have been, entirely different from the atomic bomb. I've gone up and down that Asiatic littoral, figured out how we were going to stop Soviet Bisons and Bears with their standoff missiles shooting at us from a hundred miles. It's a tough proposition to get out there and shoot them down before they can release their missiles. We've got to be able to destroy those missiles two ways: with aircraft interceptors and ship-based antimissile missiles."

On my way back east in a leave status, I stopped at General Dynamics in Fort Worth and said I wanted to fly the F-111. I'd not seen one; G.D. had only a couple flying and they were a mare's nest of difficulty. It didn't take very long and I went flying. The pilot who sat on the right-hand side was a Naval Academy graduate but he was an air force test pilot, and the plane was an F-111A owned by the government, still not delivered to the air force. After a cockpit checkout I took off in afterburner, climbed unimpressively slowly to 14,000 feet where the pilot said, "You've got to come out of afterburner, Admiral." I took the engines out of afterburner; the airplane would fly at 14,000 feet without afterburner, but it didn't fly very fast, only around 250 knots indicated airspeed. At 14,000 feet that might be 300 knots true airspeed, but that was very slow for cruise power.

He said, "That's all it will do. These are developmental engines," which was meant to say that the production engines would have more thrust. So I asked, "How high can I take it?" I wasn't going to do anything they hadn't already done, and he said, "You can take it up to about 30,000 feet in afterburner." It carried a lot of fuel because it was supposed to be a long-range machine. We got to 30,000 feet and he told me to come out of afterburner. He said, "We haven't gone much higher than this, maybe 32,000 or 33,000 feet."

I flew it around and felt out static and dynamic stability, control, maneuverability, and rolled it to the right and to the left a couple of times. I said, "Let's see how fast it will go." So we shoved it back into burner and, even though it was supposed to be supersonic by quite a bit, I only got it up to high subsonic. It was a terrible disappointment to me and actually I lost altitude in order to get the Mach number up. I started at 31,000 or 32,000 feet, ended up at 27,000. He said, "That's not where it's at its best, Admiral. Take it down on the deck."

He got permission from Fort Worth control for a large area to make a high-speed run across the flats of Texas, southwest of Fort Worth. Well, I had to admit that was pretty exciting. We got supersonic or very close to it, about 0.95 or 0.96 Mach, which is going like hell—very, very close to 610 or 614 knots on the deck, at an altitude of about 100 feet. Barns and telegraph poles were flying by pretty fast.

"Now," he said, "when the radar's in and the weapons system is in, we're going to have the capability of making a high-speed run like this and dropping bombs quite accurately."

By that time, the fuel was getting down. We must have held that speed for fifteen minutes and must have gone close to 200 miles on the run. Then we came back and made several passes around Carswell Air Force Base. Coming in to land at a nice slow approach speed, the damned nose would go up and stay up and you had to correct it. On a good airplane, it won't have much deviation but if it goes up it will nose back down, and if it goes down it will nose back up, if you keep the elevator set as you're supposed to. Not that baby! The center of gravity was already aft of the aerodynamic center, which gave you an unstable coupling. In the landing configuration, if it nosed up it continued up; if it nosed down, it went down further. You had to correct the attitude continuously in the landing configuration.

In due course the air force and General Dynamics put in electronic sensors to automatically sense and apply corrective elevator action. This was also an awful system. Nevertheless it was controllable—not wholly, but controllable for a field landing. But I knew damned good and well that it was going to be a nightmare getting on a carrier because if there's one thing you don't want to do it is to be applying up and down elevator to try to hold the nose steady in a carrier approach. You want to hit the deck right in the middle of the arresting wires. You don't want up and down deviations taking place as you aim for the deck.

Even with one flight, I had more knowledge than any other naval officer, and certainly more knowledge than any civilian leader in the navy or the government, about where that airplane stood at that particular time. Now, this wasn't the F-111B; this was the air force model, the F-111A, but remember this was a common airplane. We had great trouble with the engines. The nacelles were not matched with the engines; the engines weren't getting

"Commonality" was the Department of Defense goal that led to two General Dynamics versions of the infamous TFX—the U.S. Navy F-111B (foreground) *and the U.S. Air Force F-111A.*

enough air through them and that was the reason the engines weren't putting out the thrust and that had a lot to do with the lack of speed, rate of climb, and overall performance. It was a beast!

When I got to work as OP-05, the troops were saying, "We've got to stop that F-111B" and this, that, and the other about the airplane. It sounded awfully easy for the CNO to say, "You've got to kill it, you've got to kill it," but gosh, Congress, the SecDef, and the White House were on the line to go this way. There'd been millions appropriated for the damned airplane that later was clearly no good and should have been junked for both the air force and the navy. So I said, "I'm going to go fly the navy version; wait a while."

That was arranged and I went up to Grumman, really the first time that I'd had much to do with Grumman. The company test pilot was a fellow named Bill Miller; he had come from Patuxent and had flown the F-111B there in preliminary navy flight tests. Bill Miller left the navy to do this work and I was always sorry that he did because I thought he was a marvelous test pilot.

We did the same things in this machine that I had done in the other, and it was a sad machine. Bill was more concerned about the unstable nature of the airplane in the carrier approach configuration. It was a side-by-side cockpit fighter, with the weapons officer on the right, the pilot on the left.

The enemy could have gotten on your blind side and to have to depend on the guy on the right seat was ridiculous, difficult, and dangerous! In the F-14, you just move your head right or left and you can see the tail; all designs are going for good cockpit visibility over 360 degrees.

Well, I came away from Grumman disgusted and worried. For another year, the program went on with my constantly thinking about what could be done to fix the F-111B. The air force was not very nice to me about the airplane and my complaints. They were essentially unresponsive. I made all kinds of appeals to important people who had parts in the program, but none were willing to even admit the plane was a stinker or to talk about stopping the program. In the end, McDonnell didn't come up with anything helpful. General Dynamics was just going to Band-Aid it, as long as the Band-Aid money would last, and McNamara was determined that he was going to have a common airplane for the navy and the air force, which isn't a bad idea, but this was the wrong plane.

First of all, it was supposed to be a fighter—a highly maneuverable, high-altitude fighter for the navy and a low-altitude interdiction machine for the air force. But there's no good reason why the navy and the air force can't have a common plane. They did with the F-4 Phantom. The F-4, a navy plane, was shoved down the air force's throat.

Only Grumman came in with the answer. The Grumman people were building the tail of the F-111B and it was a big piece of business for them. They knew the carrier business well. They'd had more successful experience in it than McDonnell by quite a bit, more types of planes aboard the carriers than McDonnell has ever had. What happened was that the Grumman people came to my office one afternoon around five o'clock: Lew Evans, the president of Grumman; Mike Pelehach, their top designer; Larry Meade, another fine engineer; Joe Reese, a former navy captain who'd been working for Grumman, a damned good man. They said, "Now, do you like the looks of this?" They went over to my conference table, spread out blueprint drawings, not in a great deal of detail, but in significant detail.

It was a tandem fighter, the pilot in front, weapons officer behind him. It used a better arrangement for the engines that were in the F-111. It had the best of the weapons control systems. I had no objection to that at all, and it carried the Phoenix missile, which I also had high faith in. We had to take a lot on faith in those days because that Hughes Phoenix missile was a very ambitious, expensive step, and if it had failed it would have been a mistake of the most expensive nature—or so we thought then.

I was on knife edges from that event for the next several years. The big thing was that if we dropped the F-111 right then (if we could get rid of it), it wasn't going to cost us very much money, navy-wise. Of course, the Department of Defense would scream to high heaven. They wanted the navy to buy F-111s to keep the quantity high and improve the savings, always overwhelmed by inflation. We could take the dollars programmed for F-111Bs and let Grumman or another prime company design a good navy fighter that would do the interdiction job but would also do the dogfight job.

The Grumman proposal looked good, and the argument that we'd be ahead in the long run if we could get the right kind of a fighter was strong. We wanted the long-range, air-to-air, standoff capability. Adm. Thomas Moorer, chief of naval operations, looked at the proposal and said, "This is so much better than pursuing this F-111B. We're going to have to sell it to Secretary of the Navy Ignatius. Give me a little more time to get my ducks in a row. I'm not prepared yet to go further, but I sure like what I see, and besides Grumman knows how to build carrier airplanes a hell of a lot better than General Dynamics, who has never built one."

I went into the SecNav and put the prints on Mr. Ignatius's coffee table. He got out from behind his big desk and sat on his davenport. If looks could kill! I said, "Now, the reason I'm telling you, Mr. Ignatius, is so that you won't get caught cold, and so you can tell Mr. McNamara. Grumman came in to me with this thing. I have talked to the CNO about it and he told me to come and tell you."

I said, "This proposal is a far better fighter/interceptor. I've had enough experience in these matters and I tell you that the Naval Air Systems Command isn't at all happy with the F-111B and the program. Some of our very best people are absolutely against that plane. You are just perpetrating something that's going to boomerang on you, on Mr. McNamara, and the president, everyone. I'm not against a common navy and air force airplane such as is the F-4, but that F-111B airplane is a bad one. It's not very good for the air force as a low-altitude bomber, and it's a disaster for the navy as an interceptor and fighter. I think we ought to have a competition, but this time let the navy run their program. They know how carrier airplanes must be built, not the air force, not a company that's never been in the act."

There had been that big controversy over President Kennedy taking the TFX contract from Boeing, the competition winner, because politically he wanted to do something for Texas where his popularity was very low in the early sixties before he was assassinated. So they changed the decision. It would have been a scandal today, monkeying with this thing. It was a scandal then. But the secretary of defense was a pompous, irritating, conceited smart-ass, wrong as hell about defense matters—and many people were afraid of him. Later, President Johnson, who was from Texas, was also on the bandwagon.

So, Grumman came in and showed this front-back-seat version, and the selling point was that at that time the navy hadn't yet put much money into the F-111B; we hadn't gone to production. Grumman showed two cost curves; one curve was the projected cost of continuing with the F-111B up through the buy, which was very large. The other showed the projected cost of dropping the F-111B and starting a new fighter/interceptor, using engines, radar, and other components that were available. The curve that was to apply to what is now the F-14 crossed the F-111B curve in about three years, and did not rise as fast as or as far as the F-111B.

There were reasons for that. The F-14 was going to be a smaller airplane and not as heavy. It wasn't going to have a very expensive escape system, which the F-111A later abandoned because it wasn't adequately developed. At that time, both companies were projecting costs that were not influenced by the tremendous inflation that hit, not only the country, but particularly aerospace industries in about 1972 or so. Every year everything rose in price. Everyone expected inflation, planned for it, raised prices and wages accordingly, and of course, caused it. It was still low, around 4 percent. So it all looked pretty reasonable and I was perfectly willing that anybody should try to tear our new plan apart. I wanted them to, but no one tore it apart at all. I now fully believe that the bulk of defense analysts and officials, the Congress, and even the White House were relieved of the F-111 problem and glad to see it go away, at least in the navy where it was pure disaster.

There were hearings in the Senate that took place up at the Senate Armed Services Committee, of which Senator John Stennis was chairman. It was a hearing to see whether they would provide funds to continue with the F-111B. So we went up there for the Senate Authorization Committee hearing—Secretary of the Navy Ignatius, Chief of Naval Operations Admiral Moorer, Assistant Secretary of the Navy for R&D Bob Frosch, and myself, the deputy chief of naval operations for air. This hearing went on hour after hour after hour. Mr. Ignatius answered every question. Any question any member of the committee or staff asked, he'd answer. He was playing the game of staying with the F-111B, and we who were opposed weren't getting a chance to get in the act—not Moorer, not I, nobody. He had the knowledge; there was nothing stupid about Ignatius—not stupid, but ignorant and unwilling to do anything not originating in SecDef's office. This had been a hot potato for a long time and there was an awful lot of opinion in these areas.

Finally, about five o'clock Senator Stennis was called from the room; he came back and said, "This question is for Admiral Connolly. Admiral, you're a professional, you've had the technical training, the combat experience, and the fleet experience, and all that. Would you choose to stay with the F-111B or would you go for the new fighter?"

I said, "I'd go for the new fighter."

Then he said to me, "How would you feel about the F-111B if new engines were put in to overcome the lack of power?"

I said, "There isn't enough thrust in all Christendom to make that airplane a fighter."

That was the coup d'état, and Ignatius was fit to be tied. Then Mr. Stennis went around and said, "Now, the next one. Admiral Moorer, what is your choice?"

He said, "I understand what the secretary is saying, Mr. Chairman. I understand that and I realize the points he's made and all that, but I would support Admiral Connolly."

It was a miracle—the fastest change that happened since World War II. We got Congress to kill the F-111B funding. They stopped those funds, and then they made funds available that were already in hand for the F-111B to be used for the new airplane. Actually, McNamara gave up; I guess he was in such deep trouble across the board in so many ways—I'm now talking about the year 1967 or thereabouts. New people came aboard in January 1969, the business of running the program went back to the hands of the Naval Air Systems Command, and the right people took over and ran what turned out to be the F-14. There was a cleanup squad left to extract the navy financially, contractually, and otherwise from the F-111B contract.

The first flight of the F-14 took place in twenty-two months. It was a thrilling thing to watch this machine being built and I spent a lot of time up there on Long Island at meetings and decision sessions. It developed as they had thought about it during this earlier period. Grumman had the best contractors in the country on the team. Everybody was turned on, and it moved very fast from that session in the Senate until we had permission from Defense, funds, a contract, a contract for entirely new engines to be developed for the air force and the navy—F-100s for air force, F-401s for the navy, using a common core, a common gas generator. Pratt & Whitney won the competition. Hughes got on the team, and they made the AWG-9 and that Phoenix missile system work. There were doubting Thomases all over this town—in the Pentagon, on the Hill and in other companies! But it all went together well.

Now what did Defense do? Instead of putting money up to buy the number of planned airplanes, they began to buy small. The whole plan had been priced on buying at least seventy-two a year, only six a month; now came piddling numbers—twenty-four to thirty per year—thus making each airplane very expensive. Then *The Washington Post,* which had been in the act from the beginning, started talking about "the $20-million fighter." That permeated the Congress and anything that went wrong—we lost an F-14 on the second flight—ricocheted all over the place, as though the navy and air force hadn't lost airplanes in tests before. It wasn't all that easy, but the airplane came through. It went to the fleet and the pilots were crazy about it and they still are.

My career came to an end on 1 September 1971. When I left, every aviation program was in apple-pie shape, funded, progressing, the planes doing beautifully in the fleet; they were the best airplanes that any tactical force in the world had or could lay hands on. The F-14 was well on the way to being the best U.S fighter. The E-2C was a huge success; the EA-6B was the best ECM aircraft in the world. The A-6E was the star attack airplane in the fleet; the A-7E was moving ahead in the fleet. These were and still are great tactical air warfare airplanes. The Phoenix missile on the F-14s was and is the best tactical missile in the world, far superior to anything developed by the air force, or the Russians or anybody else.

8
The F/A-18
Strike/Fighter for the Twenty-first Century?

Vice Admiral Kent L. Lee

Kent Liston Lee was born in Florence County, South Carolina, on 28 July 1923, son of R. Irby and Hettie (Floyd) Lee. He enlisted in the U.S. Navy on 15 August 1940, entered the flight training program on 12 November 1942, and was designated naval aviator and commissioned ensign on 7 August 1943. During World War II, Admiral Lee saw combat in the Pacific with Bombing Squadron 15 and Fighting Squadron 15 on the USS Essex (CV-9), participating in numerous strikes against enemy positions and shipping at Marcus and Wake in the Marianas, the Bonins, Palau, and the Philippines.

Admiral Lee's postwar command assignments included Attack Squadron 46, Carrier Air Wing 6, USS Alamo (LSD-33), and USS Enterprise (CVAN-65), operating in Southeast Asia. In August 1970 he became director of the Office of Program Appraisal, Navy Department, followed by tours as deputy director of the Joint Strategic Target Planning Staff, and commander, Naval Air Systems Command, Washington, D.C. As a vice admiral, he held that billet until relieved on 31 October 1976 and he was officially placed on the retired list of the U.S. Navy on 1 November 1976.

Admiral Lee's decorations include the Distinguished Service Medal, the Legion of Merit with Gold Star, the Air Medal with two Gold Stars, the Navy Commendation Medal, the Presidential Unit Citation Ribbon, the Navy Unit Commendation Ribbon with two Bronze Stars, and numerous foreign decorations.

Vice Adm. Kent L. Lee, commander, Naval Air Systems Command, 1973–76.

On 31 August 1973 I became commander of the Naval Air Systems Command in Washington, D.C. I was told about all the projects and the problems. Recently, the Grumman F-14 contract, which had been a terrible problem for both the navy and Grumman for the previous three years, had been restructured. I didn't have to face that big problem on arrival at Naval Air Systems Command.

A very interesting thing came about in the fall of 1973, and it really was the genesis of the F-18. During previous tours of duty, I kept looking around for the solution to the navy's problems of having so many different types of aircraft on the hangar deck of an aircraft carrier. I was very intrigued with a visit I made to see the YF-16 and the YF-17 prototypes during a tour in Omaha. I had many long conversations with the vice president of Northrop, who was in charge of this particular project.

In the fall of 1973 the navy had a study on what was called a VFAX, an experimental (X), heavier-than-air (V), fighter (F), and attack (A) aircraft. This study came to the conclusion that it was possible to build a multipurpose aircraft as we had done during World War II. The F-14, for instance, is a single-purpose aircraft. Its radar is hard-wired, and it's good only for air-to-air intercepts. It was not designed with an air-to-ground capability. The A-7, on the other hand, has a radar that is only suitable for air-to-ground attacks. It's also hard-wired. With the coming of computers and the inertial platform, and with programmable radars, the study concluded it was possible to build a multipurpose aircraft. This aircraft, with the push of a button, would have a very good air-to-air radar and could fire missiles and guns. Then, with the push of another button, the radar would shift to the ground mode and would lock on ground targets. The inertial platform would keep the airplane positioned in space so you'd have an airplane that was equally as good as the F-14 in the air-to-air role and equally as good, if not better, than the A-7 in the attack or air-to-ground role. These items were not too expensive, fairly reliable, and they were lightweight. We hadn't tried it before, but it looked to us like the way to go.

About this same time the Congress and the Office of the Secretary of Defense (OSD) were very unhappy with the F-14 program. The navy was pushed very hard to find an alternative to the F-14. In the meantime, the air force had two airplanes, the YF-16 and YF-17, in prototype. Congress and the OSD decided that the air force should complete these prototypes and then choose and build one, and call it a lightweight fighter. The navy didn't really want a lightweight fighter; we didn't think that it would do our mission. What we liked was the VFAX.

After many days of arguing, Deputy Secretary of Defense William P. Clements, Jr., decided to let the navy issue a procurement request for bid studies for a VFAX airplane. We in Naval Air Systems Command put together a set of specifications that were really based on this VFAX study and that described the airplane we wanted. We had sent that out to industry, and in the meantime the air force was moving along to have a competition between the YF-16 and the YF-17, now called the air combat fighter, when the Congress redirected our program. The Congress said that we had to join forces with the air force and pick one of the lightweight fighters for our alternative airplane. Congress was going to severely restrict the numbers of F-14s we could buy.

At the time there happened to be a former carrier pilot by the name of Chuck Myers who was director of air warfare in the office of the deputy undersecretary of defense for tactical warfare programs. He opposed our VFAX initiative, but he was overruled by Mr. Clements and James R. Schlesinger, secretary of defense, and our VFAX was approved.

Myers had very close contact with some of the staffers on the Hill. These staffers got together with Myers, and they put language into the House and

Senate conference bill that set aside $20 million for our joining the air force's air combat fighter program. That's how I believe it came about—Chuck Myers in the Department of Defense and staffers on the Hill. There wasn't much of a way out of it. The navy then joined forces with the air force, and the debate went on for some time as to whether or not we would have to pick the same airplane.

It was decided that we could send our VFAX documents to both the YF-16 and YF-17 contractors. The YF-16 was built by General Dynamics, the YF-17 by Northrop. Since neither of these companies had ever built a carrier-suitable airplane, we also requested that they take partners. Northrop chose McDonnell Douglas for the YF-17, and General Dynamics chose LTV (Ling-Temco-Vought) for the YF-16. In addition, through these weeks of debate, it was decided that we could have separate competitions. The air force would have their competition between the YF-16 and the YF-17, and we would have our competition. The hope was that we would both settle on one airplane or the other, the YF-16 or the YF-17.

As luck would have it, the air force finished their competition and persuaded Mr. Schlesinger to let them go ahead and release their winner about a month or six weeks before we were ready. Our procurement documents were put on the street perhaps two months after the air force documents, so at the time the air force was ready to announce the winner for their selection, the contractors hadn't even finished sending in their documents for our competition.

Mr. Schlesinger decided to let the air force announce that the YF-16 was the winner of the competition. He announced—and he also sent the Congress a letter to this effect—that the navy's competition would continue, and that we would be allowed to pick whichever airplane was most satisfactory to the navy, the YF-16 or the YF-17. We continued our competition for another six weeks, perhaps two months. As it turned out, our winner was the YF-17. The YF-16 we deemed to be unsatisfactory for carrier use, and, of course, when that happened, the roof fell in. Congress was unhappy; the staffers were unhappy; the contractors were unhappy. We were going to build separate airplanes, which was true. On top of that, LTV, which had assumed the lead on their bid for the naval air combat fighter, lodged a bid protest with the General Accounting Office.

As a result, we had to go to Congress and testify, which we did for weeks on end. LTV launched a very bitter bid protest, and its primary claim was that we had no choice other than to pick the F-16, since the Congress had told us to join forces with the air force. Our argument was that we had been given permission to pick whichever was best for the navy, and the Congress had been notified and had not objected. The General Accounting Office ruled that what we had done was entirely legal in every aspect, and that the F-18 contract with McDonnell Douglas, which we had signed on the day we announced the winner, would stand and was legal.

About the same time, Mr. Clements decided we would call the F-17 derivative the F-18. There was so much controversy associated with the YF-17 and YF-16 that Mr. Clements felt that it would give a new life to the F-18 to give it a new name; just that simple. So that's how the F-18 program got launched.

I think the F-18 has turned into a marvelous airplane. The reliability and maintainability are unbelievable, as compared to other airplanes, because we worked on them very hard, from initial specifications all the way through to the production aircraft. It's a very fine fighter plane, very maneuverable, with an excellent radar and outstanding engines. It's also a very fine attack plane. The only misgiving I have about the F/A-18A is the fact that it's too short-legged. We were tied into that by having to take a derivative of the F-17. If the F/A-18A had about 2,000 or 3,000 more pounds of fuel, it would be a better airplane. Currently, it has to carry that fuel externally. I think it is just about the ideal airplane; there are not many of them built. The A-1 or AD Skyraider, I thought, was the all-time best design I ever ran across, and I think the F/A-18A is a close second with one exception, and that is fuel.

There has been an axiom that it's a lot easier to take a navy airplane and adapt it for air force use than vice versa. They are different breeds of cat, although it's not as difficult as people say. I think this business of adapting air force airplanes to navy purposes has really been put in the wrong light. When the navy *wants* to take an air force airplane and adapt it for carrier use, it does so with a minimum of fanfare—the F-86 is an example. We didn't have a fighter during the Korean War that would measure up to the F-86. Our Grumman fighters and our F7U Cutlass were just no match for it. They weren't good airplanes. We took the F-86 and made it into a carrier plane, and we called it the FJ Fury, a very fine airplane. It is a valid example, because the F-86 was not carrier-suitable. It was strictly an air force airplane, and we took it and made it carrier-suitable. We rebuilt it and put in the strength points it needed for a hook and the catapult launches. In my view, we can take almost any air force airplane that is a good airplane and make it into a carrier-suitable airplane.

Carrier suitability became a political issue with the F-111 and later on with the F-15 when a good many people said that what we ought to do is buy the F-15 and cancel the F-14. Once again, that issue of carrier suitability came up. In my view it wouldn't have been difficult to make the F-15 into a

THE F/A-18—STRIKE/FIGHTER

From the U.S. Navy's VFAX study of the early 1970s came the multipurpose McDonnell Douglas F/A-18A strike fighter, proven in combat in the Persian Gulf in 1990, and destined to become the navy's primary fighter-bomber of the early twenty-first century.

carrier-suitable airplane. I think it's a matter of "not invented here." I think it could have been done, and it wouldn't have been all that expensive, although the F-15 was not the airplane we wanted. I think the airplane we really wanted and needed was the F/A-18A, because what the F/A-18A does for us is something that's not appreciated by land-based aviators.

The deck space on an aircraft carrier is very valuable. If you have an airplane that can be both a fighter and an attack plane, you've got two for one. In my view, all the future tactical airplanes built for carriers ought to have that dual capability. It's so easy to put in inertial platforms and programmable radars with the current computer technology. All of our attack planes can have a very fine air-to-air fighter capability, and that just doubles the capability you have on an aircraft carrier of a given size. I think if we had been allowed to go our VFAX route initially, we would have had essentially the same airplane that we have with the F/A-18A except we would have had a little more fuel space. I believe that's about the only difference we would have had. For the cycle time on aircraft carriers nowadays, F/A-18As have to carry fuel tanks. A-4s were the same way. They always had to carry fuel tanks to meet carrier cycle times, and it's a perfectly satisfactory way to do business.

One of the unusual features of the F/A-18A program was that the pressure to build the F/A-18A and the VFAX came not from the navy but from the Department of Defense and from Congress. After I had looked at the VFAX and the possibilities there, I—and I believe all my assistants in the Naval Air Systems Command—became firmly convinced that the way to go was the VFAX. We concluded that we ought to buy a minimum number of F-14s and go this new route, which would be the future of naval aviation. I believe the entire group in Naval Air Systems Command, all the rear admirals and all the people who sat on the various selection committees, became firmly convinced that this was the way to go. The Naval Air Systems Command was enthusiastic about the VFAX, which later became the F/A-18A.

Now on the other side of the river, over in the Pentagon, the deputy chief of naval operations (air warfare), OP-05, and his assistants, wanted no part of the F/A-18A—or the VFAX, for that matter. They wanted to buy more F-14s and more A-7s. I believe the CNO also wanted to buy more F-14s. I don't think the CNO ever believed in the F/A-18A, but he was enough of a politician that he was not willing to buck the Department of Defense, Mr. Clements, and Mr. Schlesinger, and even the Congress. Consequently, he didn't oppose the F/A-18A, but he was not an enthusiastic supporter.

I think the OP-05 group did everything they could to avoid having an F/A-18A. But we were pulling for it in the Naval Air Systems Command, so that led to some big divisions between Naval Air Systems Command and OP-05. We were enthusiastic for it, and they wanted no part of it, which was an interesting commentary. Of course, with OP-05 against an airplane, it's almost never built because they have the ear of the CNO, and that's how things usually happen. In this particular case, the pressure came from Mr. Clements, Mr. Schlesinger, and the Congress, and we in NavAir were enthusiastically supporting them, so out came the F-18.

Grumman fought the F-18 the entire time I was at NavAir, because they could see it taking business away from them. They paid millions, I imagine, in lobbyists to fight the F-18. The A-7 people were fighting the F-18. I've never seen anything like it. Industry was arrayed against it; they had their congressmen alerted, and then there was a group of people in OP-05 who didn't like the F-18. The F-18 was a very controversial airplane for a while and, yes, we flinched. I felt we had a world beater, a world-class airplane except for fuel, and it's proved itself. The F-404 engine is one of the best engines ever built, although they did have some problems later in the program. At 800 hours one of the compressor blades goes. That's not a big problem; they'll fix that—and the radar is one of the best ever. Everything about the F-18 is good, and it's going to be around a long time.

By the summer of 1976, I'd been at Naval Air Systems Command for three years, and that's a very tough job. The reason it's so tough is that you have so many projects, so many things going on. The commander always has to take the projects that are in deep trouble and try to straighten them out, take them forward and get additional funding or whatever else is needed. The F-18 was a very wearing competition. In the summer of 1976 I decided what I'd really like to do is retire. I had thirty-six years in the navy, and also, the F-18 program had been signed, sealed, and delivered, so I put in my papers.

I think the Naval Air Systems Command was a very rewarding tour. I think we did some good things for naval aviation. I had the finest team of officers and civilians I've certainly ever seen, and I believe it was as good a team as NavAir and the old Bureau of Aeronautics ever had. I think we did some good things for naval aviation in the reliability and maintainability area and in various other projects we had going on at the time—the cruise missile, the Harpoon missile, Sidewinder, Sparrow, the CH-53 helicopter, and two dozen other projects that we worked on during this period of time.

From working as a mechanic on an SBC to helping put together the F-18, I had a great thirty-six years in the navy, and I'd do it all again.

The Washington Scene

One of the lessons of World War II was that future conflicts would have to be fought by armed forces whose separate components—army, navy, and air force—had merged into a "unified" force under some sort of central direction. The National Security Act of 1947 was signed into law by President Harry S. Truman on 26 July 1947. The law created the National Military Establishment (changed to the Department of Defense in 1949) and provided for three separate military departments that reported to a secretary of defense. These three departments were empowered to train, supply, administer, and support the land, sea, and air forces, which received their operational direction from unified (e.g., CinCLant) and specified (e.g., SAC) commanders reporting to the secretary of defense through the Joint Chiefs of Staff.

This historic legislative action followed many long months of extremely bitter wrangling by the U.S. Army, Navy, and Army Air Forces, each of which had its own view of its proper role in an unsettled postwar world. Few people in government, military or civilian, had been able to agree on what a single military establishment should be, even though most agreed that one should be created. Some wanted more centralized control, others less, and still others felt things were fine the way they were, although perhaps a little more cooperation among the services would be in order. The divisions between the three branches of the military were deep and bitter; doctrines and beliefs held to be sacred since the days of Gen. Billy Mitchell and "unsinkable" battleships, nurtured and reinforced by successes and failures in the campaigns and battles of World War II, died slowly, if at all.

The debate over "unification" was waged at a time when events on the international scene prompted reactions and policy decisions on a grand scale at home and severely tested the capabilities of the new Defense Establishment to function effectively. The Truman Doctrine, Marshall Plan, Atlantic Pact, and North Atlantic Treaty Organization (NATO) increasingly placed burdens on the U.S. military to be maintained in constant readiness and alert.

The new U.S. Air Force had a plan for a seventy-group force, with a Strategic Air Command equipped with new B-36 intercontinental bombers; the navy wanted a modern carrier fleet with marines for tactical ground support; the army wanted a broad range of new weapons to support a ground war in Europe—and President Truman wanted a balanced defense budget of $15 billion. Sincere men on all sides argued persuasively for spending more (or less) on planes, ships, and guns, and the burden of achieving a so-called "balanced force" fell to the very man who had fought "unification" vigorously—the first secretary of defense, James V. Forrestal. Forrestal eventually succumbed to the extreme pressures of the job, when, on 22 May 1949, he took his own life.

Louis Johnson became the second secretary of defense on 28 March 1949, and soon thereafter, made a controversial decision that sent the navy to "general quarters." On 23 April 1949, based on views expressed by Chief of Staff of the Air Force Gen. Hoyt Vandenburg, Chief of Staff of the Army Gen. Omar Bradley, and retired Gen. Dwight D. Eisenhower, acting as a consultant, Secretary Johnson halted construction of the navy's newest and biggest aircraft carrier, the USS *United States*. He took this action without individual consultation with Chief of Naval Operations Louis Denfeld or Secretary of the Navy John L. Sullivan, in the belief that the new carrier represented a duplication of the air force's strategic bombing mission. Secretary Sullivan resigned in protest and was replaced by Francis P. Matthews.

What followed in the wake of the cancellation was the "revolt of the admirals," an intensive effort by the navy to discredit the air force, the B-36 bomber, and Secretary Johnson. The opinion was expressed by many that the tactics used by the navy were questionable, if not irresponsible; in any event, the B-36 survived the congressional investigation that followed and the revolt came to an end.

In the aftermath of the fracas, Adm. Louis Denfeld became a scapegoat and was relieved as chief of naval operations because of his role in the controversy. Other naval officers who had played key roles in the affair survived to fight future battles over budget and force levels. Secretary Johnson went on an economy drive that soon had all of the services upset with his high-handed methods, and concerned about their state of combat readiness. The concern was soon justified when, on 25 June 1950, North Korea invaded South Korea, and the United States became involved in another war in a far-off place for which we were unprepared to fight. In September, Johnson himself assumed the role of scapegoat when he was asked by President Truman to submit his resignation.

After allowing the defense posture of the United States to decline in the years leading up to the Korean War, President Truman reversed his stance and boosted defense spending to over the $40 billion level. The new defense posture relied on the bombers and nuclear weapons of the Strategic Air Command to deter general war, with conventional forces available to fight small-scale, limited wars. President Dwight Eisenhower assumed office in January 1952 with a mandate to cut the defense budget once again, with continued reliance on the Strategic Air Command as the first line of deterrence. This "New Look" defense policy closely resembled Truman's, but had a higher price tag. Following a reorganization of the Defense Department and the end of the Korean War in 1953, the battle of the budget began in earnest, with a projected drop in the navy's FY 1955 budget of $1.5 billion from the previous fiscal year. Over one-half of the navy funds were earmarked for naval aviation, however. With the constant fight for scarce dollars, continuing questioning and redefinition of the roles and missions of the services in an era of a technological revolution in aircraft, weapons, and electronics, tensions increased between the services, and once again the battle was carried out in the news media and congressional committee rooms.

On 24 February 1956 Senator Richard B. Russell, chairman of the Senate Committee on Armed Services, appointed Senator Stuart Symington, former first secretary of the air force, as chairman of a subcommittee on the air force to study air power. With the fox loose in the chicken coop, the navy, along with the army, was soon included in the proceedings. The actions of the subcommittee dramatized the importance and difficulty of retaining air and missile superiority, in general painting a rather dismal picture of America's state of military preparedness in the fields of long-range jet bombers, continental air defense, and ICBMs and IRBMs. Although there were differences of opinion among the military and civilian witnesses and the members of the subcommittee, there was general agreement that the Soviet Union was closing the gap in air power and long-range missile capability, and that modernized, balanced army and navy forces were a necessity.

In the mid- to late fifties, all of the services were deeply involved in the missile development business. Names like Titan, Atlas, Jupiter, and Polaris grabbed the headlines and a large share of the defense budget, accompanied by the usual interservice squabbling on the sidelines over who should have the responsibility to develop and operate which type of missile. The fight for the aircraft carrier earlier in the decade finally began to pay off, as the USS *Forrestal* (CVA-59) hoisted her commission pennant on 1 October 1955, while her sister ships *Saratoga* (CVA-60), *Ranger* (CVA-61), and *Independence* (CVA-62) were under construction. These ships, along with the two fossil-fueled carriers and one nuclear-powered carrier whose keels were to be laid before the end of the decade, would form the backbone of U.S. carrier aviation through the 1980s.

This pattern of high-visibility defense-related issues, with their attendant behind-the-scenes bickering would be repeated, with an occasional aberration or variation, in the decades to follow. In an atmosphere of continuing international crises, the Cold War, and an occasional shooting war in the Western Pacific, the navy would always have its periodic fights with the air force, the secretary of defense, and some members of Congress over the aircraft carrier. The issue was not necessarily over the existence of the carrier—although the carrier was once called the world's largest floating anachronism—but how many are enough? At the root of the arguments were the convictions of air force leaders of the omnipotence of air power—in particular strategic bombing—and conviction that the proper executor of missions connected with this doctrine was, of course, the U.S. Air Force. These turf wars between the navy and the air force had their origins in the 1920s, in the days of William Moffett and Billy Mitchell, and will probably never end. A roles and missions fight always lurked just beneath the surface and would arise on occasion, particularly with the introduction of a new technology that could give one service a capability not precisely defined in current mission statements.

Frugality and cost cutting were the rallying cries of practically every new president and his secretary of defense since the end of World War II, but frequently these good intentions couldn't withstand the pressures brought to bear by the Joint Chiefs of Staff or Congress, concerned about military preparedness in the face of some unexpected event on the international scene; a Sputnik placed into orbit, a "missile" or "bomber" gap, a Korean War, or a Suez crisis could quickly raise the cap on the defense budget in dramatic fashion.

From 1950 to 1960 each secretary of defense came to his job with marching orders from the president and with his own distinct style of management and leadership (or lack thereof). Louis A. Johnson, George C. Marshall, Robert A. Lovett, Charles E. Wilson, Neil H. McElroy, and Thomas S. Gates, Jr., presided over the defense establishment during that decade, with varying degrees of success. But little of what the military or Congress had learned during those years about how to manage the defense establishment prepared them for the disturbing days ahead.

Robert S. McNamara was sworn in as President John F. Kennedy's secretary of defense on 21 January 1961, after a fourteen-year stint with Ford Motor Company and a meteoric rise from manager of the planning and financial analysis office to president of the company. McNamara brought with him to the Pentagon a troop of young, irreverent, and inquisitive "whiz kids," who applied the methods of the analytical statistician to the programming and budgeting process. A new phrase quickly became part of the military staffer's lexicon—"cost-effectiveness."

"Is it worth the cost?" was the benchmark by which a new ship, airplane, or tank was to be judged—not the judgment of military leaders with years of experience in the field. The military did not take kindly to their cavalier treatment by McNamara and his band of young inquisitors. During his term in office McNamara was accused of many things—muzzling the military, lack of respect for Congress, ignoring the nation's defense needs, and cancelling pet military projects, often without prior consultation with the service staffs. The Skybolt missile, the nuclear-powered aircraft, the Snark missile, the supersonic bomber B-70 were a few of the programs eliminated or cut back by McNamara, much to the chagrin of their service sponsors. Program cancellations were newsworthy, but they shared headlines with international crises and other defense issues of the day—the aborted attempt to land Cuban rebels at the Bay of Pigs in 1961; the Cuban missile crisis of 1962, which brought the world to the brink of nuclear war; the Cold War and the Vietnam War; and a singularly controversial defense procurement issue—the TFX joint service warplane contract.

In November 1962 the Pentagon announced that the contract for the airplane would be awarded to the General Dynamics Corporation's Convair Division in Fort Worth, Texas, despite findings by a Pentagon Source Selection Board composed of air force and navy officers that a design submitted by the Boeing Company would be better and cheaper. Damaging contradictions of McNamara's rationale in awarding this contract were brought out by senior military officers, aeronautical engineers, and procurement experts in testimony before Senator John L. McClellan's Permanent

Subcommittee on Investigations. Despite allegations of conflict of interest and other improprieties, McNamara succeeded in pushing the air force version of the airplane, the F-111A, into production, despite cost overruns and deficiencies in aircraft performance that had been predicted well beforehand. It was not McNamara's finest hour, but he survived and continued to exercise a level of arbitrary authority unprecedented in the history of the Pentagon.

McNamara's disdain for the judgment of technical experts and experienced military men came to the fore once again in 1963 when he decided that the navy's next aircraft carrier would be conventionally rather than nuclear powered. Once again a McNamara decision precipitated a congressional hearing; this time Senator John Pastore's Joint Committee on Atomic Energy listened to scores of witnesses who pointed out the significant military advantages of nuclear power for surface ships. Although the final report of the committee indicated there was no valid technical reason not to build a nuclear carrier, the committee considered McNamara's assertions that a conventional carrier would be more cost-effective were without foundation. The USS *John F. Kennedy* (CV-67) was built as a conventional carrier, the last of her kind to be commissioned in the U.S. Navy in the twentieth century.

By the time McNamara left office in 1968, a series of four reorganizations of the Department of Defense had taken place since the enactment of the National Security Act of 1947. With each reorganization, power became more and more centralized in the Office of the Secretary of Defense, in the name of greater efficiency.

Following the passage of the National Security Act, there were many in the U.S. Navy who genuinely feared for the security of the nation. Among these was Vice Adm. Gerald F. Bogan, a veteran carrier task group commander of the Pacific war. In Admiral Bogan's opinion, the confusion as to the role of the U.S. Navy had caused morale in the service to sink to its lowest level since he entered the commissioned ranks in 1916. In a remarkable letter to Secretary of the Navy Francis P. Matthews, Admiral Bogan expressed his views on this deplorable situation, asserting that the country was being sold a false bill of goods on the advantages of unification. Admiral Bogan's account of his embroilment with the Washington bureaucracy and his subsequent demotion is the story of an officer with great courage, integrity, and devotion to the navy.

Adm. Don Griffin was one of those rare officers with the perfect combination of intuition, foresight, and understanding of the issues that enabled him to argue effectively and persuasively for the U.S. Navy in the conference rooms of the Pentagon and the Congress on three separate occasions when the future of the aircraft carrier was in contention. As a special projects officer on the navy staff in the Pentagon during the "revolt of the admirals" in 1949, then-Captain Griffin was heavily involved in preparing the testimony for Chief of Naval Operations Louis Denfeld to deliver to the House Armed Services Committee. When the dust had settled, Admiral Denfeld had been fired, but other officers involved in the fight—Radford, Burke, and Griffin—survived and each attained the rank of admiral during his distinguished career.

Former Chief of Naval Operations Adm. George W. Anderson explores dimensions of Pentagon management and operations in a story noteworthy for both its candor and its restraint. His staff of handpicked senior flag officers—McDonald, Smedberg, Russell, Donaho, Pirie, Sylvester, Sharp, and Griffin—were officers whose hundreds of years of operational experience and expertise stood in marked contrast to the inexperience so prevalent among the high civilian authorities in DOD. Admiral Anderson touches on his professional and personal relationships with Robert S. McNamara as they dealt with such wide-ranging issues as assignment of flag officers, the Berlin and Cuban crises, relationships between the military and the civilians in OSD, and lastly, the TFX affair. Admiral Anderson provides a rare insight into the aftermath of that controversy and the final episodes of his turbulent term as chief of naval operations.

Although the final selection in this section falls slightly out of place chronologically, Adm. Arleigh A. Burke's unrestrained and far-reaching commentary on political and military affairs does not lend itself to the artificial constraints of time—or subject. Admiral Burke—statesman, "sailor," public servant, and philosopher—served an unprecedented three terms as chief of naval operations and his unsurpassed understanding and depth of knowledge of international affairs qualify him well to speak on practically any subject. The Vietnam War, the Suez crisis of 1956, the reorganization of the armed forces in 1958, the Bay of Pigs—the list could be endless. Above all else, however, his writings reflect an unmatched comprehension of the significance of the navy to the power of a nation and a superb ability to formulate his views on the subject and articulate them for the American people.

9

"Bickering Is Still the Rule; Unanimity Is Non-Existent"

Vice Admiral Gerald F. Bogan

Gerald Francis Bogan was born in Mackinac Island, Michigan, on 27 July 1894, son of James H. and Katherine (Nash) Bogan. He attended Lane Technical High School in Chicago before entering the U.S. Naval Academy on 15 June 1912, from which he was graduated and commissioned ensign on 3 June 1916. After eight years of assignments ashore and afloat, he reported in August 1924 to the Naval Air Station, Pensacola, Florida, where he had flight training and was designated naval aviator on 16 March 1925. Numerous squadron and ships company tours were followed by assignment to command the USS Saratoga *(CV-3), operating in the South Pacific in support of the Solomon Islands campaign. Admiral Bogan's distinguished war record also included commander, Naval Air, Tenth Fleet, commander, Fleet Air, Norfolk, Virginia, commander, Carrier Divisions 25, 11, and 4, and commander of task groups in Task Force 58/38 in the Pacific area until the end of the war.*

Following World War II, Admiral Bogan assumed command of Fleet Air, Alameda, California, and on 2 February 1946 became commander, Air Force, Atlantic Fleet, with the accompanying rank of vice admiral. He became commander, First Task Fleet on 8 January 1949 and remained in that command until his retirement, effective 1 February 1950.

*Vice Adm. Gerald F. Bogan,
commander, First Task Fleet,
1949–50.*

*In addition to the Navy Cross, Admiral Bogan's decorations include the
Distinguished Service Medal with Gold Star, the Legion of Merit, and the
Presidential Unit Citation with four stars.*

In October 1949 Secretary of the Navy Francis P. Matthews invited all flag
officers to write to him expressing their opinion of the existing morale in the
navy and the causes for it. As commander, First Task Fleet, I wrote my letter and forwarded it through the commander in chief, Pacific Fleet, Adm.
Arthur W. Radford. The subject was "Comment on a statement of Captain
John G. Crommelin," because Captain Crommelin was sounding off all over
the country, creating a great deal of publicity. He was assigned to the OP-23
division in the Navy Department in Washington. So when I wrote this letter,
I forwarded a copy to him in the Navy Department, and without my concur-
rence or even telling me about it, he arranged a secret meeting with the press
in the elevator of the Press Club in Washington and handed them this letter.
It created a great deal of publicity all over the country. The secretary's aide
at that time told me later that my letter had been on his desk for five days and
he hadn't read it and hadn't done anything about it, but when Crommelin

gave it to the press and it was published all over, the muck really hit the fan. Those were the aide's exact words.

The letter read as follows:

"My dear Mr. Secretary:

"At the beginning it is proper for me to state that in no manner have I, to date, endorsed or condemned Captain Crommelin's statement because no one has asked me to do so. Had such been the case honest necessity and conscience would have required hearty and complete agreement with the affirmations made in his release to or interview with the press.

"Your dispatch, which prompts this letter, is surprising in its interpretation of the motive in the basic statement. It avers that the Crommelin statement and subsequent public utterances have embarrassed the progress of unification and harmony and the Navy Department. It further states that these remarks have been inspired by apprehensions concerning the future of Naval Aviation. Mr. Secretary, while realizing that this is your honest belief, that interpretation of the genesis of Crommelin's release is the most superficial gloss and does not remotely touch the heart of the question. The basic reason behind all of it is a genuine fear in the navy for the security of our country if the policies followed in the Department of Defense since the National Security Act became law are not drastically changed, and soon.

"It is necessary for me to assert now to you that I opposed the act as written and passed and so testified before the committee. My reasons for opposition and suggestions for other methods of achieving ultimate unity in the Military Establishment were given at that time, 1 July 1947. I forecast much of what subsequently occurred. Records of that testimony are available. The creation of three departments or sub departments where formerly there were but two is not unification. Under the present law it can be made to and does operate effectively in the field. But it would be sheer balderdash to assume that there has been anything approaching it among the secretariat, the Joint Staff, or the high command of all three services. Knowing that honest differences of opinion must constantly be present, bickering is still the rule; unanimity is non-existent.

"The morale of the navy is lower today than at any time since I entered the commissioned ranks in 1916. Lowered morale, to some degree, may be expected to follow any war during the readjustment to the organization for peace. In my opinion this descent, almost to despondency, stems from complete confusion as to the future role of the navy and its advantages or disadvantages as a permanent career. Optimistic letters and plans issue from Washington. And concurrently the situation deteriorates with each press re-

BICKERING IS STILL THE RULE

lease. The younger men are necessarily concerned with their future security. We of greater age and, we hope, more mature judgment are fearful that the country is being, if it has not already been, sold a false bill of goods. Junior officers in large numbers, whose confidence I enjoy, have come to me asking advice on their future course of action. I have invariably encouraged them to enhance their professional ability against the day when the troublesome questions now paramount would be equally resolved. It is becoming increasingly difficult for me to do this honestly.

"If the adequate Military or Defense Establishment could be achieved without a navy or naval aviation, I would gladly advocate using funds now expended to maintain that service, on the procurement of the best other necessary weapons and equipment. Not even the United States can support indefinitely, during peace, the tragically large military budgets we are now devouring.

"There is no cheap quick victory possible between any two nations or groups of nations each having strong even if relatively unequal power. Yet at a time as potentially critical as ever existed in our history, the public has been lured into complacency by irresponsible speeches by advocates of this theory. The results could be a great national and worldwide catastrophe.

"I have been informed that when the committee hearings resume in October the navy will be afforded the opportunity to state its case completely on the items comprising the agenda. Since Captain Crommelin's press statement, I am more optimistic than before that such will be the case. But the agenda does not cover the fundamentals of our national security. It embraces a total of eight items, all pertinent but by no means the complete whole. It is my earnest hope that at some time in the near future this vital subject may be thoroughly explored with no consideration being given to the reputations nor politics of the witnesses who appear. It is bigger than personalities, broader and deeper than politics. It is our country.

Respectfully,
C. F. Bogan"

Admiral Radford forwarded my letter with this endorsement: "Admiral Bogan is an officer of great ability and wide experience of naval aviation. No question of his sincerity and high principles. I know that the writing of this letter was motivated by sincere patriotism. Rightly or wrongly, the majority of officers of the Pacific Fleet concurred with Captain Crommelin and with the ideas expressed by Vice Admiral Bogan, but most of them avoid any statements to that effect, but they would probably question the timing of such public statement. Nevertheless, it would be a great mistake to underestimate the depth and sincerity of their feelings. Because of my conviction that this letter is representative of the general feeling, I commend it to your attention."

Now, Raddy sent that to the secretary, through Chief of Naval Operations Adm. Louis Denfeld. An acknowledgment on 8 October from Secretary Matthews said: "Dear Admiral Bogan: Your communication of 20 September has been forwarded to me by the Office of the Chief of Naval Operations. As soon as time permits, I shall write you further with respect to this document and your procedure in handling it." He's got this listed "Secret and Personal." There was nothing irregular in my handling of the thing.

Now, the aftermath of that was that I was in Pearl Harbor on an operation when I got this thing. When I got back to San Diego, I had some inkling that I was going to be sent to Jacksonville in a job subordinate to the one that I'd had for three years, and I made my plans to retire if necessary. So then I got a letter from Adm. John Dale Price, who was a classmate of mine and was vice chief of naval operations, and he said there'd be no decision about Jacksonville until he came out here on a certain date. Three days before that date, I found my orders to Jacksonville, which meant going back to rear admiral after four years, so I put in a request for retirement from the First Fleet on 1 February 1950. I heard about my being demoted by reading about it in the public press, about the orders to Jacksonville.

Then Crommelin again sent this so-called newspaper reporter down from San Francisco to interview me one night. He got himself very drunk, and when I had to go across the street for some purpose, he telephoned the San Diego and Los Angeles newspapers from this house, setting me up for a press conference with them the next day, unknown to me. Next morning the telephone rang.

"What have you got to say at this press conference, Admiral?"

I said, "I don't know anything about a press conference."

He said, "We got a call from your agent last night saying there would be a press conference." It was the local press and the Los Angeles press.

I said, "There's no press conference. I'm going to attend a Republican Women's Club luncheon today, but I have no press conference scheduled or intended."

But when I got down there, there were two reporters from the San Diego papers, and they asked me if I'd give them a little background on this letter. I told them that the letter had been widely published and that Secretary Matthews had said that any letters he received in reply to his invitation would receive serious consideration. I said, "In my case, the only consideration I'm positive I received had him refer to me in very derogatory terms to

a committee of the Congress and also questioning the handling of my letter."
Well, that was fine.

So the next day I went flying, and when I came back from flying, here
was Vice Adm. Thomas L. Sprague at North Island to meet me, and he said,
"I have orders to relieve you immediately."

This was a week before I was due to retire. I was ordered to temporary
duty in the 11th Naval District, went back to rear admiral for a week, so I
would not retire with four stars with a gun deck promotion on retirement at
that time, and that was it—and Adm. Forrest Sherman, chief of naval operations, and Matthews did that together. Sprague said that he argued with
Forrest Sherman about it, but Sherman said, "The decision's been taken.
Shut up."

═══10═══
The Revolt of the Admirals

Admiral Charles Donald Griffin

Charles Donald Griffin was born in Philadelphia, Pennsylvania, on 12 January 1906 and was raised in Washington, D.C. In 1923 he entered the U.S. Naval Academy, was graduated and commissioned ensign in 1927, and in 1930 was designated a naval aviator. Subsequent service included extensive operational and staff duty ashore and afloat, in both the Atlantic and the Pacific and in Washington. In major assignments at sea Admiral Griffin commanded the aircraft carriers USS Croatan *(CVE-25) and* Oriskany *(CVA-34), Carrier Division 4, and the U.S. Seventh Fleet. Major staff assignments included duty on the Joint War Plans Committee of the Joint Chiefs of Staff; special assistant to the chairman of the Joint Chiefs of Staff; director, Long Range Objectives Group; director of the Navy's Strategic Plans Division; deputy chief of naval operations for Fleet Operations and Readiness; commander in chief, U.S. Naval Forces, Europe; and commander in chief, Allied Forces, Southern Europe, from March 1965 until relieved of active duty, pending his retirement, effective 1 February 1968.*

Admiral Griffin's decorations include the Distinguished Service Medal with Gold Star, the Bronze Star, and the Presidential Unit Citation with two Gold Stars.

Adm. C. D. Griffin, commander in chief, Allied Forces, Southern Europe, 1965–68. As a captain, Admiral Griffin played a key role in the preparation of navy position papers during the "revolt of the admirals"—the B-36 affair.

In September 1948, as a captain, I received sudden orders to proceed to Washington, D.C. and report to the Strategic Plans Division (OP-30) in the Department of the Navy (OpNav) as officer in charge of special projects. That meant practically nothing to me when I heard this. It wasn't too long after I got back there that I got head over heels into the business of the so-called revolt of the admirals—the B-36 affair. I found that the special projects had to do with the preparation of statements for the chief of naval operations on very critical points.

The first one that I got involved in was Adm. Louis Denfeld's statement to be given before the House Armed Services Committee in connection with the B-36 investigation. I worked on that for some time, prepared a draft, and put it in his office. Every day I would call his office to find out whether he had read it, whether he liked it, whether there were any changes to be made in it, and to get some guidance as to what was going to be done with it. My questions along this line were unproductive. It soon became quite apparent to me that Admiral Denfeld was not going to take any fast action on this because he, himself, was feeling his way along. Really no one knew precisely how this whole thing was going to turn out. It was being handled very carefully.

I made several trips from Washington out to Pearl Harbor to talk to Adm. Arthur Radford, CinCPac, on this particular subject, because if there was any one individual who had kicked this thing off, it was he. My job was really to make sure that there was perfect liaison between Admiral Radford's headquarters and Admiral Denfeld's headquarters. This was extremely interesting. I don't think I'll ever forget the traumatic experiences we had at that time. This was the time when OP-23 was established under Arleigh Burke. This agency existed for the principal purpose of gathering information that could be of value to the case of the navy in convincing the Congress and the people of the United States of the function of a navy—why we must have a navy, what a navy does. This was in the general background of cutting back on the total defense effort. Expenditures for military purposes had to be cut back very sharply. It's quite obvious that no chief of service is going to willingly permit his own service to be cut back, and allow that money to be spent for another service. This was really a question of life or death for the navy.

One of the big issues at that time was the tremendous sums that were being spent for the B-36 bomber. Finally, it came to a point of gloves-off and bare-knuckle fighting, and a lot of internal fighting. This was publicized very greatly in the press, not all of it completely accurately, as is usually the case. It made good headlines.

The day before Admiral Denfeld was to appear on the stand before the House Armed Services Committee I got a call about seven o'clock in the morning to come to his office at eight o'clock. When I arrived, there was Ozzie Colclough, who later on was the Judge Advocate General of the navy, and then the Dean of the Law School at George Washington University; and two other people whose names I can't remember. We four were given the job of drafting Admiral Denfeld's statement. We took my statement as a rough draft and worked on that. We worked on it all day long, eating lunch and dinner in Admiral Denfeld's office. We were joined very briefly after dinner by Arleigh Burke, Bob Dennison, who at that time was naval aide to President Truman, Admiral Radford, and Admiral Denfeld.

The last page came out of the typewriter and was approved by Admiral Denfeld at three o'clock the following morning. He delivered the statement at ten o'clock that morning before the Armed Services Committee, and Secretary of the Navy Matthews was just wild. I use that word deliberately,

In the late 1940s the Consolidated B-36 strategic bomber became the focal point of bitter interservice arguments involving the future of the carrier navy.

because he had been trying for days to get a copy of Admiral Denfeld's statement for the committee. He had been told there was no statement, and it wasn't available. He didn't get the statement until just about an hour before it was to be made to the committee, and by that time, he was on his way over to the committee, so he had no chance to read it. He sat there and listened to it being given by Admiral Denfeld.

When Admiral Denfeld was about three-quarters of the way through his statement, the secretary got up and walked out of the room, obviously very displeased. Admiral Denfeld's statement was received with great approval by the Armed Services Committee. He was congratulated on the forthrightness of it. I might say that it was received with practically unanimous approval by the naval service as a whole. Practically everyone felt that it was a very courageous statement for him to make, and one that had to be made.

Later on that afternoon I was in Admiral Denfeld's office with Rear Adm. Stuart Ingersoll, who at that time was OP-30 (Strategic Plans), when the telephone rang. It was Bob Dennison on the other end of the phone from the White House telling Admiral Denfeld that a statement had just appeared on the press ticker at the White House that he, Denfeld, had been fired as chief of naval operations.

He took it like a man but he was, no question about it, hurt that the president had seen fit to fire him without even calling him and telling him so in person. The other people who were involved in it didn't all get hurt. I didn't get hurt and it was well known, I think, that I had a lot to do with writing that statement. Arleigh Burke didn't get hurt; he was under a shadow for awhile, but he later was made the chief of naval operations. And Admiral Radford was not hurt.

There's no question about it that Admiral Denfeld was the one who was hurt. He was a very fine man, absolutely wonderful in personnel matters. In the infighting in the Joint Chiefs of Staff, this was something a bit foreign to his nature to get in there and be involved. It took an awful lot of courage for him, considering the nature of the man, to deliver his statement, because it was a statement in opposition to his own secretary, in which things were said about the other services that sometimes are not said. As many people said, somebody had to tell the story and who best to tell it than the CNO. He did it very courageously knowing full well that he was laying his career on the line. But he did it.

═══ 11 ═══
Crisis, Controversy, and Deception

Admiral George W. Anderson

George Whelan Anderson, Jr., was born on 15 December 1906 in Brooklyn, New York, son of George W. and Clara (Green) Anderson. He attended Brooklyn Preparatory School, and entered the U.S. Naval Academy in 1923. Graduated and commissioned ensign on 2 June 1927, he subsequently advanced in rank to admiral when he took the oath of office as chief of naval operations on 1 August 1961.

Following graduation from the Naval Academy, Admiral Anderson held a number of increasingly important aviation billets afloat and ashore. He was in the aviation detachments of light cruisers, a carrier pilot flying from the old Lexington *(CV-2) and the* Yorktown *(CV-5), patrol plane pilot, and navigator in the second* Yorktown *(CV-10) under the famous Capt. J. J. "Jocko" Clark. Ashore he served in important positions in the Plans Division of the Bureau of Aeronautics during the early part of World War II and later on the staff of Vice Adm. John Towers, ComAirPac. As the war was winding down, Admiral Anderson was aviation officer in the Strategic Plans Section of the CominCh staff in Washington. He was later commanding officer of the escort carrier* Mindoro *(CVE-120) and the fleet carrier* Franklin D. Roosevelt *(CVA-42), and subsequently served with distinction in various staff assignments afloat and ashore. In September 1959 he became com-*

ADMIRAL GEORGE W. ANDERSON

Adm. George W. Anderson, chief of naval operations, 1961–63.

mander, Sixth Fleet, and commander, Naval Striking and Support Forces, Southern Europe, with the accompanying rank of vice admiral.

Admiral Anderson served as chief of naval operations until relieved of all active duty pending his retirement, effective 1 August 1963. He was sworn in as ambassador to Portugal by Secretary of State Dean Rusk on 4 September 1963, serving in that capacity until 1 July 1966.

In addition to the Distinguished Service Medal with Gold Star, the Legion of Merit, the Bronze Star, the Commendation Ribbon with Combat "V" (Navy), the Commendation Ribbon (Army), and the Presidential Unit Citation Ribbon with one star, Admiral Anderson has been awarded numerous foreign decorations and awards.

I was sworn in as chief of naval operations on 1 August 1961, relieving Adm. Arleigh Burke, who had enjoyed a very memorable tour of duty for six years as chief of naval operations. The principal responsibility of the chief of naval operations is as a member of the Joint Chiefs of Staff, and here we had Gen.

CRISIS, CONTROVERSY, AND DECEPTION

Lyman Lemnitzer as the chairman of the Joint Chiefs of Staff. We had Gen. Curtis LeMay as the chief of staff of the air force, Gen. David Shoup as the commandant of the Marine Corps, and Gen. George Decker as the chief of staff of the army. I found them very easy to get along with. They were strong for their own services, which one would expect, and if there were differences of opinion, in my own opinion, there always should be among members of the Joint Chiefs of Staff. They involved either the readiness provisions for their own services or else a different interpretation of what was best for the country in any particular situation.

One of the things that had always disturbed me through the years was the controversy and a lot of infighting that had prevailed between the air force and the navy, focused particularly on naval aviation but, after the air force became separate, on the navy as a whole. I was determined that I was going to try to minimize unnecessary conflict and controversy between the services. This came to a head, to me, almost immediately because I was told that Secretary of Defense Robert McNamara had expected me, as the chief of naval operations, to keep a rein on General LeMay and the air force. I sent word back that I was going to have no part of that, I would call my shots exactly as I saw them, and I would support the air force when the air force was in the right. It was up to them to justify their own programs, just as I had to justify the programs for the navy, and I didn't expect them to be throwing harpoons at particular navy programs that I felt were necessary for the readiness of the navy.

McNamara's people thought they would exploit the long controversy that had prevailed between the navy and the air force, and I just rejected that approach. I think, under the circumstances, my relationship with General LeMay, who was the chief of staff of the air force during my two years as CNO, was excellent. Actually, we had differences of opinion at times. He didn't always agree with me on certain naval aviation programs but, basically, the hostility that had prevailed for a long time was greatly diminished, and I thought that was a good accomplishment.

The first major international event that occurred was the so-called Berlin crisis. That was more or less under way when I took office. Now, the Joint Chiefs of Staff generally met three times a week. On Monday afternoons they met for a short period of time. The secretary of defense, his deputy, and frequently his deputy for international affairs, then Paul Nitze, would come down and meet with the Chiefs. The second meeting was on Wednesdays, and the third meeting was on Fridays. Those were the routine meetings of the Joint Chiefs of Staff. In the interims, the deputies for plans and operations would meet and go over the agendas in advance, settling as many as they could agree on or that did not require the personal involvement of the Chiefs. For each of these meetings, we would have a Joint Chiefs of Staff briefing for the service chiefs, a briefing by their own staffs—air force and army, I know, also did it. We'd go over each matter on the agenda in addition to reading the papers. We had the oral arguments within our own organization as to what position I would take or, if I was not present, the vice chief would take or the deputy for plans, Admiral Sharp, would take. So I kept quite familiar with all the matters of the Joint Chiefs of Staff, to the best of my ability.

In any event, this Berlin crisis was building up and, having followed the previous crisis, which was the Bay of Pigs in Cuba while Admiral Burke was CNO, the civilian authorities became deeply involved in this, and took it with a degree of, let me call it, hysteria. McNamara's approach, having been in contact with McGeorge Bundy at the National Security Council level, and, I guess, reflecting President Kennedy's views, was that he had to make some public decision every week to demonstrate the solidarity and the firmness of the determination of the United States to the Russians. So there had been announcements of certain deployments and redeployments of forces, and many of these were items that were included in a long list of things that might be done by the different services and by the Chiefs to indicate a stronger position by the United States.

One of the items on the list was the transferring from the reserves to active duty of some fourteen ships and fifty aircraft squadrons, I believe. From the navy point of view, these were things that should be done immediately in the event that hostilities were imminent or hostilities had occurred but which were very unwise to do at that time because, when you call up reserves, you sort of eliminate that group of reserve officers from the available pool. It's sort of a one-shot affair because you have an immediate impact on the reserves, unless you're continuing to expand and mobilize.

So I was upset and told the secretary of defense I was opposed to the calling up of these reserve aircraft squadrons and destroyers. Nevertheless, it was directed, and, in retrospect, in subsequent years, we saw the adverse effect of that, because to take people from their civilian life and employment and just retain them on active duty for an indefinite period of time—and that happened for a year—just pulled the rug out from our whole reserve program.

Well, eventually the Berlin crisis sort of faded away and in the case of the navy, we were "stuck" with these aircraft squadrons and reserve ships on active duty and the problem of what we were going to be able to do to build up our reserves again. You could not demobilize the reserves because that would indicate that you no longer had this position of national strength that had been undertaken in a series of operations. I believe strongly that you've got to depend on the active duty forces of the navy to meet any contingency that develops, up to the point that hostilities are imminent.

One of the steps that can be taken—and it's a good step—is putting a temporary freeze on the demobilization or discharge of people at the end of their enlistments or reserve officers or officers resigning or being transferred from active duty to reserves. By holding them in, you retain the people of experience and then, when that particular crisis is over, let them go, and by that time you've had more time to effect replacements. If there is a crisis, while it may disturb the personal plans of some individuals, they normally recognize that this is part of their service and their duty to their country, much more so than recalled reserves if there is not a true perception of a crisis to the country, because that really interferes with the lives of the individuals on a continuing basis.

The other matter that was always of concern to the Joint Chiefs of Staff even then and achieved much greater emphasis later on during the Vietnam War was the concept of gradualism, or gradual approach. I believe that when military forces are committed by the United States—and the United States should be reluctant to commit—they should be committed in adequate force on a basis of winning. Nothing other than winning, once military forces are committed, should be acceptable. I recognize that there can always be a set limited objective to a particular problem, but to utilize that as a concept of war is wrong.

The principles of war have been pretty well proven over the years. They've been enunciated quite well in the service war colleges and I think should be adhered to. With this new group of so-called political scientists, if you want to include McNamara as one—certainly he was surrounded by them—I think there was an error in their approach to military problems. I feel, also, that it would be far better to have the Joint Chiefs of Staff more intimately related to their commander in chief, the president, in more frequent contact with the president—directly or through the chairman of the Joint Chiefs of Staff—in an effort to minimize the demands on people's time that prevailed at the time when I was a member of the Chiefs. I think President Kennedy was more dominated by the secretary of defense and by McGeorge Bundy, who was his national security adviser at the time. I think it would have been much better if there had been more frequent and direct contact between the chairman of the Joint Chiefs of Staff and the president.

I'd been in office several months and trying to do the best job I could (a) as a member of the Joint Chiefs of Staff, and (b) to enhance the proper relationships of the uniformed services with the secretary of defense, which I had observed were very unfortunate under McNamara. So, one day, I suggested to the Chiefs that we ask for a meeting with the secretary of defense in what we could call executive session, and we would talk very frankly with him—tell him what we thought was wrong, what could be done to correct it and improve the relationships between the military and the civilian side of the house. The Chiefs all agreed with this, so General Lemnitzer, the chairman, said to McNamara that the Chiefs wanted to meet him in executive session, and I guess that Lemnitzer probably, very properly, told him that they wanted to express some criticism or whatever it was. So McNamara came down accompanied by Roswell Gilpatric, the undersecretary, and Lemnitzer said the Chiefs wanted this meeting at the instigation of George Anderson so he suggested George speak first.

Well, that caught me a little bit off base, so I said, "I have three items that I think are susceptible to correction. The first is that we are being inundated with demands from your office on matters that are basically the responsibility of the chief of naval operations—requests that go down to various levels in the CNO's staff, which take up a great deal of time to prepare answers—and further questions come down, further demands for detailed information. These requests are being posed by one of your staff officers who cannot even pass a U.S. naval security clearance. As a matter of fact, his name is Deter Schwebs and he served in the German Air Force during World War II and has had no experience with the U.S. Navy, particularly naval aviation, in which he's particularly interested.

"The second area in which I'm concerned is the activities of your public relations side of the Office of the Secretary of Defense, in which too much detailed supervision is being exercised by your assistant secretary for public relations, Arthur Sylvester, who made a remark that he'd fixed Admiral Burke and he was going to see that the navy was not going to have any more of these orientation cruises that they arranged to proselytize people. That program is a good program, it serves the interests of the country, and I object to the attitude taken by Arthur Sylvester in this regard.

"This third thing is one that is upsetting, and I just pass it on for what it's worth. That is one of your principal staff officers, who also cannot pass a navy security clearance, was asked at a dinner party the other night what he did and his comment to his dinner partner was that he made and broke flag and general officers."

McNamara said, "Give me his name and he'll be out of here in twenty-four hours."

I said, "Well, Mr. Secretary, I didn't anticipate this, but since you ask me his name, his name is Adam Yarmolinsky."

CRISIS, CONTROVERSY, AND DECEPTION

McNamara turned sort of white and Gilpatric reacted immediately. He said, "I can't believe that, because Yarmolinsky is too smart to make a remark like that."

The net result was that they went around the table and LeMay talked a little bit about strategic forces, the army didn't have much to say, and Dave Shoup saw the handwriting on the wall and kept pretty quiet at that point. So the net result of it was that I'd put the secretary of defense's staff on report, so McNamara goes up to his office, calls all these people in, and says the chief of naval operations has put them on report!

When I assumed duty as the chief of naval operations, I decided to set very simple objectives for the navy as a whole for the first year I was CNO. The simple objective that I set for my first year was to make everything we had work. These were the pieces of equipment that we had on hand, the weapons, the systems, the airplanes, the ships—do everything we could to put the resources into those systems so that they would work to their full potential. The first of these really focused on the guided missile programs of the navy and the surface-to-air missile systems, which were not performing as well as had been hoped for. So we had arranged to have a project manager appointed to handle the three Ts—the Terrier, the Tartar, and the Talos, the three principal surface-to-air missile systems. From the initial examination, it was clear that a lot of money would have to go into the picture in order to make all these weapons reliable. With less emphasis, we had similar attention given to the various electronic systems—the radars and the computers to make everything we had work.

My objective for the second year was "everybody is important," to emphasize the importance of people in the service, operationally, in technical areas, afloat and ashore and in staff work. Of course, this was partially directed toward making the civilian authorities in the Office of the Secretary of Defense recognize that these were smart, dedicated, capable people that we had and they had enduring responsibilities, whereas the staff officers in the OSD were anonymous people who came and went from their various think tanks in industry. Our people were career people; they were trying to do their very best and they held responsibilities.

The other major problem that I inherited was the question of the TFX, the development of a single fighter that would serve for the air force, the navy, and the Marine Corps. This was a project started by McNamara. The idea was commendable, provided it was properly studied, carefully evaluated, and a proper decision was made. One could not fight the idea any longer of trying to achieve such an objective. Unfortunately, the requirements of the services were somewhat extreme and beyond a reasonable prospect of being satisfied. So, one of the first efforts was to reduce the requirements, which were dictated primarily by the ability of the aircraft to operate effectively from an aircraft carrier deck and still to do the job.

We had over the years established a concept of standardization. This was a proven practice as well as a good concept—the standardization of fittings, piping, electronics, wiring, lights, bulbs, all of those things. We made a great deal of progress under the standardization committee with the British during the war. However, this was an old term, standardization, and that did not suit the whiz kids of the McNamara program. They wanted a different term, so they introduced a thing called "commonality." Commonality went far beyond the realm of practicability. In other words, you had to have everything in an airplane the same, and when you went that route, then you impinged greatly on the ability to satisfy the requirements of the services. That's where the lack of operational knowledge came in.

But in any event, airplane companies that were involved were submitting proposals—Grumman, North American, Boeing, and General Dynamics—which would come up for evaluation, but always one would be less undesirable than the others, which the OSD people would bat around as the most favorable. The studies went on and on, and the Joint Evaluation Board consistently made its judgment. The final evaluation came up and I signed it off with General LeMay; it then went to the service secretaries, Fred Korth for the navy and Eugene Zuckert for the air force, recommending that the better of the two was the proposal submitted by Boeing, in contrast to one submitted by General Dynamics, who had by that time gotten Grumman involved with their proposal. I knew that Grumman, an extremely experienced designer and producer of naval carrier aircraft, had endeavored, but not been wholly successful, to have its views prevail with General Dynamics, the senior partner of the affair.

So the recommendation was that the Boeing submission was the better of the two, technically and costwise. We turned it in and I attended one briefing, I think, in the secretary of defense's office, which showed what the result of the evaluation board was. I heard no more of it until one day Secretary Korth called me in and said, "George, I know you'll be happy to know that General Dynamics and Grumman's proposal has been adopted by the secretary of defense." I said, "What? Thank God for Grumman." That was my remark, which he interpreted as full approval, which it was not.

I later found out that General William F. McKee—LeMay being out of town—was called by Secretary of the Air Force Zuckert, and told that General Dynamics was chosen. McKee told me he asked Zuckert, "Who made the decision?"

Zuckert said, "I did," and Bozo McKee said, in his jocular way, "Secretary, you didn't make that decision. They don't let you make even the little decisions, much less ones involving billions of dollars." In any event, the word went out.

During this whole period of evaluation, I had cautioned the navy people who were privy to this information not to leak any information on this matter because with a multibillion-dollar contract potential, somebody might think they could make a fast buck and I didn't want anything like that to happen. When Korth told me what the decision was, I sent for Pirie and said, "Bob, what the hell can they do with this airplane if they produce it? It isn't satisfactory for the purpose intended." Bob's reply was, "Perhaps they could use it as a strategic bomber, which is what they're now proposing—the F-111, with a lengthened fuselage, to be a new strategic bomber."

But I said, "The decision has been made. You've got to do the very best you can to make this as acceptable a plane for the navy, for carrier use, as it's possible to get. Don't fight the problem any more, but try to make this as good an airplane as it can be."

That was basically what was being done, except this began to involve a considerable degree of politics, because on the one side you had LBJ, who was vice president, from Texas, and Korth, the navy secretary from Fort Worth, and the General Dynamics plant in Texas. On the other hand, you had Senator Henry M. Jackson and Senator Warren G. Magnuson from the state of Washington, where Boeing is. Boeing knew that they had a better proposal and they convinced Jackson in this regard and started doing everything they could to have this thing reexamined.

So Jackson and the Boeing people managed to have this matter made the subject of a Senate investigation. The contract had not yet been signed by McNamara. The announcement came in November 1962 that the contract was to be awarded. So Senator John L. McClellan, who was the head of the Senate investigating committee, got word to McNamara to defer signing any contract until they could be investigated. McNamara reacted to this as an encroachment on his executive responsibility, and signed the contract, which made McClellan, Jackson, and the Boeing people even more furious. Also, in their nosing around, they knew that McNamara didn't know very much about what he had signed, other than the fact that he, in his wisdom, had decided that this was the better plane and that it would be built at the General Dynamics plant in Texas.

I suspect, although I don't know, considering the relationship that Johnson and Korth had in this, that there was a lot of Texas political pressure. It was also implied that Gilpatric had some relationship with the Long Island end of things.

So, first of all, the concept was a commendable one but impractical. Secondly, they disregarded the approved procedures of the Joint Evaluation Board. Thirdly, they violated the recommendations made by the chief of staff of the air force and the chief of naval operations and, lastly, without referring back to the military chiefs that they were going to be overruled, they arbitrarily made the decision and didn't have a good justification for making it. The decision they made was for the poorer airplane at a higher cost, and this led to the prolonged investigation.

McNamara had taken the position that this was the most important decision he would ever make in his life and he must be proven right. The decision led to a compartmentation on the civilian side of those who were involved in favoring the decision and those on the uniformed side who tried to keep clear of any political controversy. I had carefully instructed all the people in the navy—and had reason to believe that LeMay had also done so with the air force—to try to stop any leaks from within the navy or any outside fighting from the navy against this whole thing, or to make it a case of civilian authority over the military. But, between the press and the Congress and the political forces that were involved, that was an almost impossible suggestion.

I was given various instructions by Korth as to what I should say. My response was, "I don't intend to say anything. If I'm called upon to testify, I will testify."

He said, "Well, you have to submit a statement of what you will say."

I said, "I don't intend to make a statement. I will answer their questions."

The civilians then made a very foolish move; they sent word down to me, in writing, what I should put in a statement, which was supporting their decisions. I gave this to Adm. Claude Ricketts, who went up to Korth and said, "Look, first of all, Admiral Anderson doesn't intend to make a statement. Second, he would not incorporate these things you've directed him to incorporate in any statement he made. Third, if he did, the first question to be posed, if he used such a statement, would be to Admiral Anderson, 'Were you put under any pressure to make such a statement?' Then, Mr. Secretary, you would have an entirely different investigation on your hands."

Well, they withdrew that request. The only thing I was told was that, when I did go up to testify, my statements would have to be given to the secretary of defense twenty-four hours in advance of my testimony.

At that time, through my executive assistant, then-Capt. Isaac Kidd, we were pretty well informed of what was going on in the Senate investigating committee. This was in March 1963. We were well prepared with documentation on the whole situation. We also knew that extraordinary measures were being taken by the Office of the Secretary of Defense for electronic eavesdropping and personal surveillance of what was going on in my office. It was like the defense in a criminal case being carried on by McNamara and his staff, defending their position, and trying to put the onus back on the military. It was a closed group, with practically no communication on the subject between us. They didn't trust us and I certainly didn't trust them.

At one point a proposal was made for LeMay and me to sign jointly a statement to the chairman of the Senate investigating committee recommending that, in the interest of national security, they drop the investigation, because it would bring out things disadvantageous to the security of the United States. I went to LeMay's office and read this thing, and I told LeMay that under no circumstances would I sign any such thing. "We didn't start this thing and I'm not about to try to end it that way."

McClellan sent word to me that he was going to call me as a witness. I said that I would be prepared to come up at any time. He said, "Be prepared to come up on very short notice. McNamara has loaded all the most embarrassing questions that they can pose to you with Senator [Edmund] Muskie to embarrass you. So, if you are prepared to come up on short notice, I will let you know when I know that Muskie is going to be out of town, and you come up and testify." Recognizing the directive that I had to submit a statement twenty-four hours in advance to the secretary of defense, I had a very careful, very firm statement prepared.

When I got the word that I was to testify the following day, precisely twenty-four hours in advance, I had the statement sent out from my office to the Office of the Secretary of Defense, time and date marked on the letter, no indication that it was urgent, in the routine mail system in the Office of the Secretary of Defense and had it logged in. This was done by Captain Kidd. Of course, I was literally complying with the directive I had, but in spirit I was not, because he had all these sharp lawyers up there who were going to tear any statement apart, but it didn't get to them in time so they were unprepared to conduct their unscrupulous type of attack on me.

I went up at the time prescribed, accompanied by Adm. William A. Schoech and one or two technical advisers, and gave as forthright a testimony as I could. Previous to this time, when Korth was trying to get me to tell him what I was going to say when I went up, I said, "I haven't made up my mind what I'm going to say."

He said, "Well, you know the facts."

I said "Yes."

And he said, "The facts won't change."

I said, "No, Mr. Secretary, but the adjectives can."

So, if I say so myself, I went up with a well-documented case, I put on the most articulate presentation I could, and any questions that were posed by the committee were answered frankly, succinctly, correctly, and very convincingly.

Prior to that time, about two weeks before, I got a call asking me if I would appear on "Meet the Press," the Sunday morning program. I said I would, and I notified the proper authorities that I was going to appear. Of course, "Meet the Press" is listed in the Sunday morning paper. I went up, prepared to meet the press, and they asked me a question about the TFX. I said, "I am not going to comment on that because any comments I have should properly be made if I am interrogated by the Senate investigating committee, and I'm not going to talk publicly on this at this time." Monday morning I got a telephone call from President Kennedy in the White House, commending me on my fine appearance on "Meet the Press."

Between my testimony in March 1963 on the TFX, and some controversial responses to other questions posed to me under oath by either the Senate or the House on previous budget matters, it was quite obvious to me that the honeymoon, if there ever was one, between McNamara and Anderson was over. Also, I guess it was apparent that the navy did not think very much of Secretary of the Navy Korth. The Senate committee knew what was going on in the Office of the Secretary of Defense. They had followed all the testimony, they had indications that Gilpatric had been close to the borderline of perjury in his testimony in response to questions, but just kept keeping the matter open for subsequent evaluation.

So a report did not come out from the Senate McClellan committee for a long, long time. McClellan was a pretty wise old fellow, and also he was a strong Democratic politician, so he had to balance things pretty well, and he deliberately kept the evaluation open to see what was actually going to happen as the contract went on, and see what the performance of the first airplanes was.

Eventually, the report came out some time later, which completely substantiated the navy's position, not only on that, but the best testimony they had was in a deposition that I had made in regard to the award of a contract for helicopters to Bell, a separate investigation. This showed that the Department of Defense operated illogically and improperly in awarding the contract to Bell, against the air force and navy evaluation. The report pretty well set forth the principles that were involved, which were those set forth in the deposition by the chief of naval operations. So, all these things together made it very apparent—and I had been tipped off on various occasions—that I was in trouble with the secretary of defense.

I was invited down to attend the annual Navy League convention in Puerto Rico and I was very meticulous in preparing my talk to the Navy League. It was basically on the subject of integrity, emphasizing the integrity of the people in uniform. During the course of this speech, Secretary of the Navy Korth was sitting there and my public relations officer came in the room and gave him a message that he had a telephone call from Washington. Korth wanted to know where there was a telephone, so Captain Thomas, my public relations officer, took him out and showed him the telephone. He didn't eavesdrop but he heard Korth say, "Yes, yes, I understand who he is, yes. Yes, I will."

He came back, looked in the room, and did not come back and resume his seat, but he wanted to know where he could find a telephone available to make some special telephone calls. Did the district have one? Rear Adm. Allen Smith, commander, Tenth Naval District, had his office there, so Korth went up there, where there were normal black telephones and also a red telephone. He said, "How about this telephone? Can I get so and so on this one?" "Yes, you can."

So he picked up the red telephone. Well, the red telephone was not a classified telephone; it was just an on-line telephone that was a navy command network. But, unbeknownst to Korth, who was trying to call Adm. David McDonald in London, every time the red telephone is lifted any place in the world in these principal headquarters, a flash goes on in flag plot, the situation room, in the Navy Department and the officer of the watch is instructed to monitor the telephone call.

Well, he got the word that the secretary of the navy was telling McDonald in London to get back to Washington, to come in civilian clothes, not to let the chief of naval operations know he was coming. So, when the message was delivered, the CNO duty officer called his boss, who was OP-03, Adm. Don Griffin. Then Admiral Griffin ran in to Ike Kidd, who was in my office, and he called me in Puerto Rico, and immediately I knew what was going on, because we had all the other sources of information over many, many months. It had been coming in to Captain Kidd from various sources, not naval officers or reserve naval officers, but people who just could not condone the type of machinations that were going on in connection with the TFX investigation and the investigation by McNamara and the Office of the Secretary of Defense.

Kidd put all this thing together and he called me back, and within half an hour of Korth's telephone call, I knew everything that was going on, and that he and Gilpatric were coming around to my house on Sunday to inform me that I would either be fired or not be reappointed. I knew about this in advance, down in Puerto Rico.

During the course of the rest of that afternoon, I gave Korth two opportunities to say something to me; he didn't say a goddamned word. He just said something had come up and he had to get back to Washington right away. He left, and I stayed on for the final evening of activities, flew back to Washington—this was on a Saturday morning—and got together with my people. Knowing that they were coming around on Sunday morning, we considered the desirability of getting to the president first. We weighed the thing, even thought of having a statement made to give to the president, but I decided not to do that at that point.

So, I got a telephone call saying that the secretary of the navy and the deputy secretary of defense would like to come around to my quarters on Sunday afternoon. I was all set to receive them. I greeted them at the door very cordially, took them out on the porch, and we sat down in comfortable chairs. Then I said to Korth, "I understand that you've come to tell me I'm fired."

Korth almost passed out, but Gilpatric recovered very quickly; he's very smooth. Gilpatric had always been a pretty good friend of mine and still is. He recovered very quickly and said, "Well, George, it's not exactly that. The president has asked me to come and inform you that he has decided that he will not reappoint you as chief of naval operations. He also asked me to extend to you the request that you become the ambassador to Portugal."

I said, "In other words, McNamara has recommended to the president that I be fired. The president has not accepted that and has decided not to reappoint me."

He said, "Well, I guess you could put it that way."

I said, "Well, I will have to think about this question of whether I will accept appointment as ambassador to Portugal."

He said, "You'll have to let me know right away. Also, it will not be possible for you to see the president until you let the president know what your decision is."

I said, "I understand that." I knew that McNamara had arranged for a trip out of Washington to go to the Pacific, and I said, "I will let you know."

Then I turned around and said, "There are a few things that I'd like to speak about on this whole thing. I have endeavored to be absolutely correct in my position, and the position of the Navy Department, of not challenging civilian authorities when they make a decision that is their legal prerogative, even though that decision may have been an unwise decision. But I really object to the way that they've been handling this affair, not laying the facts right on the table, the cards face up."

As I ushered them to the door, I concluded, "I'll let you know what my decision is first thing in the morning."

I thought it all over and decided, first, there could be a serious morale situation in the navy if I just caused a ruckus, which was not good for the navy and would be very bad for the country. The second thing was I had to do things with dignity for the good of the navy as well as for the country. The third was the point where the president of the United States requests you to do something, whether you should with propriety refuse it, which could be interpreted as pettiness or something of that sort, or whether to accept an invitation of the president to serve.

So, the next morning I called Gilpatric and said, "Please let the president know that I accept appointment as ambassador to Portugal."

Things had not broken at that point in the press. McDonald had come over from London; actually, he came from Turkey, and he was met by Claude Ricketts, who explained everything to him. So I knew about that. McDonald knew what the situation was, and so the announcement was made, that McDonald was going to be appointed chief of naval operations, relieving Anderson when his term was up. Then Anderson was going as an ambassador—no announcement was made as to where, because they had to get clearance in advance.

I was going to a *Time* dinner in New York for all the people who had appeared on covers of *Time*, and while I was up there I met Adm. Arthur Radford. I told him what had happened, and he said, "Don't do it; don't accept it. Fight it out."

"Well," I said, "no, Admiral, by my concept I just can't do it that way. I think you've got to let time go by to prove whether I was wrong or whether I was right."

Then, from New York, I was going to the Army War College in Carlisle, Pennsylvania, to give a lecture. By that time, the announcement had been made that I was not going to be reappointed—still no announcement of where I was going, what ambassadorship. McNamara was out in Honolulu; he, I knew, had strongly opposed my being given anything in government, particularly anything in the United States. The president had made the decision and he asked me to go to Portugal. Later, I got the word that McNamara had put it up to the president, "Either Anderson has to go or I have to go." Also, I got word back later that Joe Kennedy, who was Jack's father, told him he'd made the wrong choice. Also, the word had come down to me not to get rid of my uniforms.

To follow up on that, when McNamara came back from his trip, the announcement had been made that I was going to Portugal. Late one afternoon, he sent for me and I went up to his office. He was in his shirtsleeves and was coming out with a group of people there from his staff, and he looked over and said, "Hello, George," and stuck his hand out.

I said, "Shake hands with you? Hell, no."

"Oh," he said. "Come in, come in."

We sat down. We had a long conversation and I got on the subject of integrity. He said, "I have integrity."

I said, "Mr. Secretary, your idea of integrity is so far removed from anything we are brought up with in the military. It's the difference between night and day."

I went over some of these things, and he started crying. This was early in May. I also knew that he'd tried to get me out right away, which the president had not accepted.

I said, "I'm going to be around here until the first of August, and it's not good to have a feud going on between the CNO and the secretary of defense because we've both got a job we have to do. Look. I'm going to be here, and I think it's in the best interests of both of us that we carry on as amicably and harmoniously as we can for the rest of the time that I'm here. I don't intend to make any statements to the press and I'm not going to write any books."

He sort of looked relieved. I said, "All right, let's shake hands on this and forget this thing."

I walked back to my office and, on the top of my desk, saw a bunch of press statements made by Sylvester and people in the Office of Public Information in the Department of Defense. "Of course, Anderson was fired. He didn't know how to run the navy very well and there were disagreements. He messed up the Cuban missile crisis," or words to this effect. There were several press statements passed out by the Office of Public Affairs in the Department of Defense.

I picked up the direct telephone to McNamara, and I said, "Mr. Secretary, any suggestion that I am not going to make any statement to the press or write any articles is no longer valid."

He said, "What do you mean?"

I said, "Well, these statements made by people from your office serve as a basis for me to repudiate any statement that I might have made that I would not make a statement."

He said, "What do you want me to do?"

I said, "Tell your goddamned henchmen to keep their mouths shut."

"Do you want me to make a statement?"

I said, "No, tell your goddamned henchmen to make no further statements on this matter. I intend to make no further statements."

In about half an hour, down the corridor comes Adam Yarmolinsky into Ike Kidd's office. He said, "I'm sorry as hell. This wasn't intended. It wasn't in the game plan!"

So, we went on and did the best we could to tidy things up and keep the ship on an even keel for the remaining few months. I have seen McNamara only once, casually, since then.

The final episode was that the president had decided that he would have a ceremony in the Rose Garden, in which he was to award me the Distinguished Service Medal, which upset McNamara terribly. Finally it was agreed that the citation for the DSM would include no mention of the Cuban missile crisis, and also it was set to be such and such a date, I think it was the thirty-first of July, and McNamara was going to leave Washington so he wouldn't be there at that time. The president changed the schedule and had it one day earlier, so he was there.

I left, went on leave, and knew that I was going to be called to testify before one of these so-called disarmament, antinuclear treaties. But also I had been invited to speak at the National Press Club. This was to occur right after I left office, after the first of August, and I said I wouldn't but I would accept an invitation to speak in September. So, I'd written a very careful address, which I had checked out by two of the best Washington correspondents. One of them, Mark Watson of the *Baltimore Sun*, who was sort of the dean of the Pentagon correspondents, said that he had certain criteria that any statement I made had to meet. First, it could not be in any way against the president of the United States. Second, it had to be specific. Third, it had to be constructive. Then he read the thing and said, "Your statement meets every one of my criteria. Go ahead and give it."

Then I called up Secretary of State Dean Rusk, and told him that I was giving a talk at the Press Club and was due to be sworn in as ambassador to Portugal ahead of time, and perhaps he might want to postpone my swearing in. He said, "Well, are you breaking with the administration, are you saying anything against the president of the United States?"

I said, "No."

He said, "We'll swear you in just as scheduled."

So, before I went up, I had two copies of my speech to the Press Club prepared, one of which I sent to President Kennedy and the other to President Eisenhower, who was then retired up in Gettysburg. I went down to the Press Club, was just getting ready to walk in, and copies of my speech, as they usually are, were ready to be handed out. I got a telephone call from the naval aide to the president, Capt. Taz Shepherd, who said, "The president's just read your speech and he asked is there any way you can keep from delivering it?"

I said "No."

He said, "Well, he says, it's liable to cause more trouble than the TFX investigation."

I said, "Well, it's too late. Just tell the president that I sent it up to him well in advance and I'm sorry he didn't get it sooner, but I'm going to give it as is."

So I gave it and it was well received. They asked questions and I gave straightforward replies to the questions—nothing dirty, as far as the replies to the questions were concerned. One of the OSD public affairs people who was there was squirming and, as he went out, there was a sign that the speaker of the day was the ambassador-designate to Portugal, and he made the remark, "Ex-designated-ambassador to Portugal."

The civilian hierarchy in OSD were all upset. They got on the White House circuit and they wanted the appointment canceled. They got the word back, the president stands by his appointment.

So, just before I left to go to Portugal, I had an appointment with President Kennedy. I spent about an hour with him talking about Portugal. As we got up to leave, I said, "Mr. President, you know, one of these days when you're out of office and I'm out of office, I'll tell you exactly why some of the things led to my speech at the Press Club."

He put his hand on my shoulder and said, "Admiral, I think I understand."

Then, we went through the period right after that when he was killed and Johnson came in. Then, later, Johnson found out that McNamara and Gilpatric had been playing footsie with Bobby Kennedy, then Johnson replaced McNamara as secretary of defense with Clark Clifford; Gilpatric had previously been eased out, too. So that was the end of that particular episode and I've only seen McNamara once since, just at a big reception. It was formal and cool and nothing was said.

=12=
A Sailor's Views on
National Security

Admiral Arleigh A. Burke

Arleigh A. Burke was born in Boulder, Colorado, on 19 October 1901. On 7 June 1923 he was graduated from the U.S. Naval Academy, commissioned ensign in the U.S. Navy, and married to Roberta Gorsuch of Washington, D.C. He served in battleships and destroyers, had various staff assignments ashore, and received his Master of Science (Engineering) at the University of Michigan. Subsequent service included extensive command and staff duties ashore and afloat as he advanced to the rank of admiral, effective 17 August 1955, when he became chief of naval operations, serving in that position until his retirement on 1 August 1961.

For the first two years of World War II, he was assigned to the Naval Gun Factory, and in 1943 went to the South Pacific where he earned the Navy Cross for extraordinary heroism as commander of Destroyer Squadron 23, the "Little Beavers," in actions against the enemy in the northern Solomon Islands area. He reported in March 1944 as chief of staff to commander, Fast Carrier Task Force 58, Vice Adm. Marc Mitscher, and participated in numerous naval engagements until June 1945.

At the outbreak of the Korean War he was ordered as deputy chief of staff to commander, Naval Forces, Far East and later assumed command of Cruiser Division 5. In July 1951 he was made a member of the United

Adm. Arleigh A. Burke, chief of naval operations, 1955–61.

Nations Truce Delegation to negotiate with the Communists for a military armistice in Korea. In August 1955 he was appointed chief of naval operations, in which position he served for six years.

Admiral Burke's numerous awards include the Navy Cross, the Distinguished Service Medal with two Gold Stars, the Legion of Merit with two Gold Stars and Oak Leaf Cluster (Army), the Silver Star, the Purple Heart, the Presidential Unit Citation Ribbon with three stars, the Navy Unit Commendation Ribbon, and many foreign awards and decorations.

While I was chief of naval operations and a member of the Joint Chiefs of Staff, President Eisenhower wanted unanimous opinion from his Joint Chiefs because it was most distressing to him when we came up with what we called split papers—differing views by the services on subjects. The differences of opinion were nearly always on missions, budgets, or controls—who does it, who can do it best? Although he was insistent on that, he had a lot of split papers, because we had some very strong-minded Joint Chiefs. The percentage was very low overall, but the Joint Chiefs happened to have a tremendous number of papers; the majority of them we did agree on, but many of those were comparatively unimportant. But on the really important papers, there was perhaps a 10 percent split, and this was quite a high percentage.

I talked to President Eisenhower quite a few times, and I believe that the differences of opinion were sound. I was a member of the Joint Chiefs of Staff because I was brought up in the navy. I knew the navy capabilities and limitations, and I had a different concept of doing things than the other services. We believed in mobility; we didn't have clear lines of demarcation between commanders. We depended upon commanders mutually supporting one another and doing their utmost to support. We didn't believe it was necessary to place a common commander over everybody and have the common commander direct all actions in detail. We believe he could call for assistance—ask another commander and get it. It works this way in the navy very well. I also thought that the air force chief was brought up the same way. They'd had certain things that through a lifetime of experience they had learned from air combat, and it was different. I'd served quite a few periods with the army at various times in my life and I knew a little bit about the army, but the army was completely different from either the air force or the navy.

The result was that every time we sent up a split paper, the president was very much concerned about it. I tried to explain why I thought that it was good that he had to make the decision, because he was the only man responsible for the whole United States. I thought this because our expertise was in different fields, but with common interests, so that the president had to make the decision on where to spend the government's money, where to put the emphasis. He had to have the unbiased direct advice of his military people to do this. He was a military man himself, but he was another one of those people who leaned over very much not to be a military man in the White House. He was less a military man in the White House than most any other president, because he didn't want to bring his military knowledge and methods of doing things into civilian government. He was right on that, but I never could convince him that he really should make those decisions. They could not be made by the Joint Chiefs unless there were some weak Chiefs, unless there was back scratching among the Joint Chiefs.

He said more than once that if the Joint Chiefs can't make up their minds jointly on all the problems that confront the military security of this country, then somebody else will make the decisions who doesn't know nearly as much about the military requirements as the Chiefs do. As long as he sat in that chair, he was the one who had to make the decisions and he did know, so it wasn't a personal matter. He could do this, but he didn't like this system.

President Eisenhower was constantly encouraging the Chiefs to go ahead with development of conventional weapons, in contrast with the nuclear weapons. I think the president realized more than anybody else in the whole world the awesomeness of nuclear warfare, and although it was possible, if it came about there would be a nation destroyed or nearly destroyed. Our nuclear power was very valuable in keeping the USSR from aggression and nothing else would have probably worked at that time. If it hadn't been for nuclear power, it's very likely that the Soviet Union would have been able to take over more of Western Europe. The air force particularly thought that they could do everything with just nuclear power. They wanted to just threaten to destroy a nation and have the ability to destroy a nation, no matter what they did. Nations would not commit aggression and would do things that the United States felt—and other "free world" nations felt—should be done.

The president knew that wasn't true. Just as John Foster Dulles had said in one of his speeches, you've got to make the punishment fit the crime, and if the Soviet Union took a small piece of territory, how big a piece would it have to be before you'd launch? They recognized the power, the limitations of that kind of power. That was all or nothing—nuclear power.

So the president said, "We've got to have a military force that can handle any situation. And that means, in a small situation, we've got to have the proper equipment and plans to correct it, and it doesn't mean that we will have to launch for everything." This was countering the air force philosophy at that time.

We had nuclear weapons in carriers. During this transitional period we were fighting hard to get the kinds of nuclear weapons that would be effective, the smaller ones, which we could handle. During that time we developed Polaris, which was only nuclear and could only be used for big events—in the category of the ultimate weapon. So he kept stressing that you've got to have conventional weapons too. Of course, I believed that and so did the army. The air force did not, in general. Because the army and the navy were trying to get the lesser weapons—torpedoes, better guns, better electronic gear, better surface-to-air missiles, using conventional explosives—most of his effort was on convincing the air force that they should have tactical weapons, tactical support.

For example, he suggested that the air force should have bombers capable of carrying conventional weapons. The air force didn't think so. But

A SAILOR'S VIEWS ON NATIONAL SECURITY

The Joints Chiefs of Staff in session. (Left to right): *Gen. Lyman L. Lemnitzer, chief of staff, U.S. Army; Adm. Arleigh A. Burke, chief of naval operations; Gen. Nathan F. Twining, USAF, chairman of the Joint Chiefs of Staff; Gen. Thomas D. White, chief of staff, U.S. Air Force; Gen. David M. Shoup, commandant, U.S. Marine Corps.*

after he was out of the presidency, in the Vietnam War, the B-52s, which were built solely for delivery of nuclear weapons, were modified to carry conventional high-explosive weapons. This is what he had in mind.

We also had some difficulty in the cost of weapons; even at that time weapons systems were getting to be so sophisticated that they were becoming more expensive and consequently you couldn't get as many as you probably would need. So we were all endeavoring to get something that would do and wouldn't cost so much. Well, this is a very hard thing to determine. In the Vietnam War, at the beginning, we could have used some World War II airplanes very, very well. They cost maybe one-tenth as much as the planes did at the beginning of Vietnam, at that time. Manufacturers weren't tooled up to make them; they would have to start all over again. But those World War II aircraft would not have done the job against modern aircraft; if they had to fight modern aircraft they wouldn't have done it at all. It's a question of keeping a mix between inexpensive and very sophisticated weapons.

Another area that the navy was heavily involved in was in the development of guided missiles prior to the development of the Polaris itself. In 1953 we did not have surface-to-surface missiles, except what really was a converted airplane—Regulus I. We had converted a submarine to carry the Regulus I; it was an unmanned aircraft in effect. What we needed mostly at that time was a surface-to-air missile to shoot down airplanes that might attack the fleet. So we developed a family of close-in surface-to-air missiles, then medium missiles, the Terrier—it wasn't a very accurate weapon or a very reliable one—then the long-range missile, the Talos. They were all similar, but they could not be used on the same mount; some of the components were interchangeable. We were trying to develop a single missile that would do all the jobs, still would not be too expensive, and be accurate and reliable. We spent a lot of money on this, and it was very hard getting that money; I had to defend this with the Bureau of the Budget or the president, and before Congress.

But the president was not very missile-minded in the beginning, although I think by the time he left office he was. He could see the need for the surface-to-air missile all right but he couldn't understand why it cost so much. As a matter of fact it did cost a tremendous amount, but you cannot invent on schedule nor can you invent on a specific amount. We did make a great many improvements, and we did develop them so they were put on as the main armament on a great many frigates and destroyers during that time.

Now, all the services had several air-to-air missiles, none of them very good, and I heard indirectly of a man out in Inyokern, California, who was trying to develop an air-to-air missile on his own—Dr. W. B. McLean. I found he was doing this in quite a bit of his spare time and also with a little government time. He needed some money, so we gave him a little bit of money, and he developed the Sidewinder, a very good, simple, air-to-air missile. This showed that sometimes something very important can be developed by a single man who has an idea, without any support at all. The development and production cost very little, and it's a very good missile still.

Regulus II was coming along and in that we had hopes for a good surface-to-surface missile, but it took a lot of space and quite a while to prepare it to fire. It wasn't something that you could shoot in ten seconds. It cost a great deal of money, and the air force and the army and a lot of other people opposed Regulus II. Finally they cut it out of the budget, which was a tremendous mistake—not that Regulus II would ever have amounted to anything as such, but if we had kept on surface-to-surface missiles, we could have developed something, a successor to Regulus II that would have been an effective weapon.

The Russians did this with the Styx. As a matter of fact, they took a cheap Regulus II and made a very good missile out of it, and we would probably have done the same thing. We are still right now trying to develop a surface-to-surface missile that's equal to the Styx. This just shows that a weapons system that can be conceived should be worked on, particularly if there's a void such as there is against surface ships—how are you going to sink enemy ships? A surface-to-surface guided missile will do it. That should have been enough for us to do research on it, but we didn't do it. You simply can't break off and resume at a later date. You lose years, and you lose the people who are interested in it; they go to some other job, so you have to get a whole new team to develop it.

In 1957, when Neil H. McElroy succeeded Charles E. Wilson as secretary of defense, the Soviets launched their first man-made satellite. Secretary McElroy took note of this whole thing and tried to put more emphasis on basic research in that area. The navy got involved in it, but we lost the battle of launching space vehicles from sea. We wanted to do that because we thought we could build a good sea-launching platform for medium-sized missiles, which could be moved around to take advantage of launching from various places. But we didn't get very much money for it. We shifted then to satellite work, putting things in the vehicles rather than trying to launch the vehicle. The ground-launching system cost so very much that there wasn't going to be any money left over for other launching systems.

In the light of what has been said about missiles and research work under way, not only in the navy but elsewhere, I would like to comment on the charge that came up in the later presidential campaign about a missile gap. The missile gap was caused by an interpretation of intelligence. At that time every service had its own technical intelligence. There was the Central Intelligence Agency, which gathered a great deal of information, but mostly political information, and each service had its own very good intelligence system. But there was great difficulty in evaluating the fragments of information that each of the services got as to what the Russians were doing. The air force always overestimated the number of missiles that the Soviets had. Army took a good deal of the same data, and they made estimates that were always on the low side, the lowest of the three. The navy's estimates were usually in between but closer to the army's than to the air force's, but it was impossible to tell exactly what the Russians were doing, and they were all estimates.

But the politicians—the Democrats at that time—took the air force data and indicated that there was a great big missile gap, and they stressed this.

Senator Stuart Symington was the leader of this thing. Actually, as it turned out, all of the estimates that we had before then were high. The Soviets were even farther behind than any of us thought they were, and within a very few months after the Democrats won office, they found that there was no missile gap at all. When it came time for the missile crisis in Cuba, the president knew very well that the Soviets had few missiles in existence, so he could act with complete assurance that it was not possible for the Soviets to launch an all-out attack against the United States. So the missile gap was a political fiction, based on interpreting all one way.

This illustrates the advantage of having the three services with differences of opinion. It turned out none of us were very accurate. Making plans based on one estimate, other than political plans, is a very dangerous thing. Now they have just one DOD intelligence system, and the answer they get is the only answer they have, and they don't have any checks on it. No matter how hard a man tries, he's got to have different interpretations of fragmentary information. You can't get it exact.

I was an old hand at dealing with proposals for reorganization of the Defense Department, having been involved originally in the Truman administration. The subject came up again in the Eisenhower administration in 1958, when the president himself proposed changes, and there was a measure finally enacted by the Congress. I opposed that all the way, because I felt that the president could use authority that he already had to reorganize the Defense Department without requiring a lot more authority that later could be used to make a single service or single chief of staff concept come true. I never quite understood his purpose, but I think he wanted a mandate from Congress to go ahead and do what he wanted to do.

There's always a tendency in a big organization to want to run it, to want to centralize it, and this will work, up to a certain size. But when an organization gets so big, not all of it can be directed from the top. The more you centralize an organization, the more you cut down the initiative of the subordinates, and eventually it ends up with the top people doing all the thinking and ordering, and the junior people carrying out orders without any authority themselves to initiate anything to carry it through.

If you centralize the services into one big service, with so many different facets, people lose their knowledge of the individual facets, and you end up with too much emphasis on one aspect of war and not enough emphasis on all the aspects of war. You're bound to make mistakes; you don't have the checks and balances that you would have if you had independent services. There's such a thing as being too independent, too; there's got to be a compromise. You've got to reach a happy medium where people take independent action, but in coordination with other people who are working within the same policy framework as you.

So I opposed this thing from the beginning, but it didn't do very much good, although the president in reorganizing only did one thing that was seriously harmful to the navy—he took command away from the chief of naval operations and put the command in the Joint Chiefs of Staff. The president insisted on taking command of the fleets away from CNO in this reorganization, primarily because of his army background and because neither the army nor the air force chiefs of staff commanded their forces. I had plenty of opportunity to express my views, which I did on many occasions.

I had the backing of the secretary of the navy in the position I took; the secretary of defense was not very strong either way. This increased the authority of the secretary of defense and he naturally liked it, because every individual, if he's good—and they were all good—feels, "If only I could make those decisions, they would be better." Well, the trouble was, as it turned out, that he didn't have time to make all those decisions, and I don't think they were better. Maybe it's all right. Any organization can be run if you have the right people to run it. But you can never get enough right people to run it as smoothly as it should be run.

I had an opportunity to serve for a couple of years longer under this change. I feel that it truly detracted from the role of the CNO. For example in Quemoy and Matsu, and the Lebanon crisis, I could give direct orders to the fleet very fast—and did. Then I went down and explained it to the Chiefs, and they, I think, always agreed that I'd done the right thing. In effect it was a fait accompli anyhow, and I could explain it to them. But they all knew about it just as soon as it happened, so that I didn't catch anybody by surprise. After the reorganization, I went through the procedure of going down and explaining the thing in detail to the Joint Chiefs of Staff, because they were responsible and wanted to make sure that they were not doing the wrong thing. Before that, if I did the wrong thing, I was responsible, not them, and it was my neck in the noose, and they didn't have to be so careful. But later they had to check and be very careful. Well, it meant that it took a long, long time to get things done, and lots of times, there was just a long discussion on what I thought was a very minute point not worth the discussion.

Then, too, because we had all been Chiefs under the old system, it didn't change very much. Thereafter I sent out dispatches to the fleet commanders, when I wanted them to do something, and say, "I intend to take up with the Joint Chiefs of Staff the following orders," and they could prepare for them. They knew what I would want to do and usually I could get it through the

Joint Chiefs of Staff. But it wasn't quite as serious as it turned out to be later on when there were new people.

Several component parts of the navy were in jeopardy during this reorganization in 1958, one of them being the status of the Marine Corps. Many army officers felt that the Marine Corps was the navy's army, and that they should have just one army, and the Marine Corps should not be in existence, in spite of the marvelous record in World War II in the Pacific. It didn't matter about the record or the individuals—theoretically it looked like they ought to have just one army, and they said that if the marines had been in the army they would have done just as well as they did with the navy. Of course, this is not true, because they're trained differently. They're trained for amphibious warfare, for small unit warfare, a lot of other things, and also they were in competition with the army and a little competition is good, too. The marines had to do better because they always have done better and they intend to keep it that way. It was an exclusive, elitist group.

Although it was a difficult time, the marines were not in nearly as much danger of losing their identity then as they were under President Truman. They came close under President Truman. But we could point to the law, and all the arguments that we'd gone through before, and we had more support on the Hill for a Marine Corps than we did under President Truman, so that battle, although it was serious, we felt we could win. President Eisenhower never made a statement that the Marine Corps should be part of the army; I never heard him. I think he kept his hands off of that pretty much.

We also had trouble with our aviation in 1958. They wanted to cut our carriers, both the other services did—the air force more than the army—but still they both wanted to make drastic cuts. This was a question of money—carriers cost money, airplanes cost money. We proved, first by our history, what we had been able to accomplish that nobody else could have accomplished, and we also proved to Congress that the only way we could do many jobs was with mobile air—carry our air bases with us. The air force simply could not do many of the jobs with land-based air; just as in Vietnam, they could not possibly have done that job with only land-based air. They need naval air. But over and over again, the air force felt that we were duplicating their mission, that there was only one group that should be responsible for aircraft and that should be them.

Actually I think it would be much better if the army had more of its own air, although it has a great deal more now than it did then. But they have to have a pretty sizable air force, and the navy has to have a pretty sizable air detachment to fulfill the missions of the United States, not just the missions of their service.

A SAILOR'S VIEWS ON NATIONAL SECURITY

But what we lost in 1958 was not so much the loss in the individual services as control of the individual services—control of the departments shifted from the secretaries of navy, army, and air force to the secretary of defense, and from the chiefs of staff of the various services to the Joint Chiefs of Staff. So now all important tactical and strategic problems are settled by committee or by the secretary of defense, instead of by the people who certainly should be more competent than the people in the other services. There was a very great downgrading then of the individual secretaries, and more centralization in the Department of Defense, and I don't think that's improved any of the services at all.

There are two things that I never talk about. One of them is women, because I don't know enough about women, and the other one is the Bay of Pigs, because I know too much about the Bay of Pigs. Now, the reason for that is that President Kennedy required all reports of the Bay of Pigs to be submitted to him, and they're gone.

There was a complete breakdown of communications channels, not only in the Bay of Pigs but every other way. There was a change of plans at the last minute. There were conflicting orders given to different people. There was an unreasonable amount of secrecy involved, so that people who should have known about the operation didn't know about it. There was not enough checking by anybody including the Chiefs. The Chiefs themselves did not realize how little the administration knew or how small their capability was for that kind of thing, and we didn't insist upon knowing. They would have told us probably, but we were not tough enough. Our big fault was standing in awe of the presidency instead of pounding the table and demanding and being real rough. We set down our case and then we shut up and that was a mistake.

Now, the trouble is that there were a lot of orders given that I don't know anything about. That operation was not under the military. We were told every time we got anywhere near it that we had no responsibility for it; we were not supposed to comment on things, unless we were asked to. It was not our show; it was a CIA operation and we stayed the hell out of it, and they would not permit any regular force of the United States to become involved in this.

The Chiefs didn't know about the operation until after John F. Kennedy was president; the planning had not been under way for a long time. This is one of the sad things about it—the planning. The first time I knew about that was when our navy intelligence people uncovered some operations in Guatemala, so I knew something was cooking. It was either just before the inauguration in January that we were informed that there was going to be an operation, or just after the inauguration, but in any case, we were not briefed on the operation; we were given the rough plan, which was really just a synopsis of an operating plan, to comment on. That's when we told the president, in writing with our endorsement on it, that we could not comment on the operation because it had no logistics annex and no communication annex and that sort of stuff, and we'd have to know that situation before we could make any intelligent comment. But from the looks of the operating plan, we thought it had about a 50 percent chance of success, from Trinidad.

It came very close to being successful, that operation—very close. A couple of little things could have happened or if a couple of things had not happened, it might have been a successful operation, it was so close.

One of the things I realized even before I became chief of naval operations was that it's very difficult for a civilian or for another service to understand just what significance the navy is to the power of the nation. Nations have lost their existence because they didn't understand the importance of seapower. Nations that understood it have lost because they didn't know how to use seapower. Control of the sea, and the maintenance of sea communications, is a necessity for an important nation. It's a necessity for any nation, but for a small nation, other people can control it and let them in on it.

Now, after I realized that this was a big gap in our educational requirements, that we had to educate people with real truth, research, real examples, and there couldn't be any just emotional froth in it, I established a section called OP-09D, which was given the responsibility for developing the basis for explaining seapower to everybody, including the navy. Now, what is it? Why is it that seapower is so important? Is it really important? Haven't things changed over the generations?

It's true that control of the Mediterranean has always been a necessity for any Mediterranean power to become a great power of the Mediterranean. Turkey was a powerful nation as long as she had a navy, and when she lost it she lost everything. Now, did the loss of the navy come first or did she lose something else and then lose the navy? Britain was powerful for centuries, when she had control of the sea. China in the very early days was most important when her junks plied to the Middle East, and then when they didn't do that, she became a landlocked nation. Germany lost two world wars because she never understood how to control the sea. She came awfully close to it, but she didn't grasp the necessity for the continued action of the slow throttling of a nation by blockade, by taking a nation's merchant ships little by little. There's no one spectacular action that causes countries to lose their navy, it's a whole series of consecutive actions. Every time it costs them so much that they can't bear the cost eventually.

Well, that lesson has been lost in history over and over again, and it's lost right now. Now, you don't have to have a tremendously large navy, I mean an overwhelmingly large one. What you have to have primarily are the naval people who can use their initiative, to carry out the policy of their government, both in peace and in war. This means training people within the navy. But first it means you've got to have a policy, a governmental policy to use not only the navy but all the powers that a nation must have and must use to remain powerful in order to have an influence on future events in the world.

Well, that means the people in the navy had to understand it. So OP-09D put out information within the navy—on showing the flag, on the importance of the navy under various conditions, why you had to have a navy before you could have any expedition overseas, because you couldn't support any operation overseas for any significant length of time without control of the seas. Why was it that the Soviet Union did not have to control the seas? Because she is a land power; she only had to deny us control. Why was it that we had to have carriers until we could find some other way of protecting our sea lines? Not just to fly airplanes over a nation, but in order to protect the merchant ships, which was the backbone of the strategy. Why was it, if you wanted to land some place you had to have an amphibious force? So you can land against the will of the people who are there, if that becomes necessary. Why were submarines so important and why, if you put ballistic missiles at sea, were they invulnerable? Why did they become invulnerable at sea, at least until somebody got an antisubmarine capability—which nobody now has?

No nation in history became a world power until it gained control of the sea. This is what Russia found out in the Cuban missile crisis.

Khrushchev knew well that the USSR didn't have the missile power compared to the United States, so if it came to a showdown, and there was an exchange of missiles, the Soviet Union would have been destroyed and the United States would not have been badly hurt. He knew that we knew it, too. So I don't think he had any idea of a nuclear confrontation—ever. In spite of having all the studies that the Soviets had made, and all the words that they had spoken about the importance of seapower, he never realized the importance of power at sea. How helpless he was when he was confronted with one American destroyer that said, "Turn this ship around"; there wasn't a damn thing he could do. He could storm and rave and threaten, but if that destroyer had the guts to make it happen, the destroyer would either sink the ship or the ship would turn around.

Now, this is what really jolted Khrushchev, I think. So the Soviets decided then, in 1962, that they had to have a nuclear capability, which they were desperately working on to equal or better ours. But also that was the first time they ever, in spite of all their studying and reading of the effects of war, really realized what a navy was all about. So they started; by 1966 they had developed a good navy, and this is why they are now putting such tremendous emphasis on it.

Seapower is composed of many different things. One of them is knowledge—of currents, ships, sea bottoms, sonar conditions—in other words, sea knowledge. The next one is fish. Fish are very important and getting more important, and not only are the fish important but the fishermen are important—they know where the fish are and they can learn a tremendous number of things about that. Next, the merchantmen are the base for the whole thing; merchant ships carry the goods, and 95 percent of the goods of the world are carried by ships, maybe more. In peacetime we form relationships with people to whom we ship. A nation forms commercial relationships; they're not as important as some people have said they are, but they're of some importance.

In any case, merchant ships plying into a nation's harbor learn an awful lot about the nation. A nation can become dependent upon another nation's merchant marine, and this Russia knows—that is straight economics among other things. So she developed a merchant marine first. There's another reason for developing a merchant marine first and that's because she had not had any real high-seas sailors for a long, long time, and you help develop those by sending your people to sea in merchant ships, just so they can learn that the ocean is big and powerful.

Then, of course, there's the navy, that is supposed to support our merchantmen and our fishermen, supposed to do research and support other oceanographic research. The U.S. Navy has now forgotten that one of its missions is to support the merchant marine; so has the merchant marine forgotten that mission. They went different ways, and the navy knows very little about our merchant marine now or the importance of a merchant marine, and the merchant marine knows very little about our navy now. The same thing is true of fishermen.

Having said all that, I think you can't just fight all-out for a navy. There is a need for a good army and a good air force, too. You've got to have some sort of balance, and everybody's idea of what that balance is will be different.

CRISES, CONFLICT, AND LIMITED WAR

In the summer of 1946, Adm. Marc Mitscher, commander in chief, Atlantic Fleet, his chief of staff, Commo. Arleigh A. Burke, Vice Adm. Forrest P. Sherman, deputy chief of naval operations for Plans and Operations, and his assistant, Capt. George W. Anderson, visited Europe to confer with military leaders and to see firsthand the horrible devastation that prevailed in Germany. Although Admiral Mitscher returned early with appendicitis, the three future chiefs of naval operations—Sherman, Burke, and Anderson—continued on the tour, which included a visit to Naples, Italy, the headquarters of the commander of the U.S. naval forces in Europe.

The visit would have great postwar significance, for, throughout the rest of the trip, Admiral Sherman formulated in his mind an expanded role for the U.S. Navy in the Mediterranean. When he arrived back in Washington, he gained approval for his plan from Chief of Naval Operations Adm. Chester Nimitz and the secretary of the navy. As a result, the U.S. Sixth Fleet was established in the Mediterranean—and Admiral Sherman was appointed as its first commander.

The assignment of two carriers to the Mediterranean, on duty at all times with national and NATO responsibilities, became an irrevocable commitment in the decades to come. However, as much as the Communist threat dominated American strategic planning and force structure, the NATO alliance was only one of a series of alliances that committed U.S. armed forces to the defense of other countries all over the world. The ANZUS Treaty (1951) allied the United States, Australia, and New Zealand, while the Southeast Asia Treaty (SEATO) (1954) united the United Kingdom, France, Thailand, Pakistan, and the Philippines to "meet the common danger. . . ." Similarly, the Rio Treaty (1947) committed the United States and twenty Latin American nations to mutual defense, while bilateral agreements with the Philippines, Nationalist China, South Korea, and Japan added to the burden that more often than not was shouldered by the U.S. Navy.

By 1960, the United States was committed to the defense of some forty-five sovereign nations, in addition to its own territories. Other countries would fall under the umbrella of such unilateral proclamations as the Eisenhower Doctrine, wherein President Dwight D. Eisenhower promised intervention in any Middle Eastern state that was invaded, if such action were requested by the lawful government.

Carrying out treaty commitments, safeguarding U.S national interests, protecting endangered U.S. citizens overseas, or responding to provocative threats or Communist acts of aggression—in hundreds of such instances, plus one "police action" and one "limited war"—the United States relied on military force to accomplish its objectives. It could be argued that the carrier battle groups were critical to the success of both crisis management and combat operations, and have consistently been the most used and most successful of U.S. general-purpose forces in peacetime and in war.

On 27 June 1950 (Far Eastern Time) President Harry S. Truman ordered naval and air support of the Republic of South Korea's effort against a North Korean invasion begun two days earlier. The outbreak of the war caught U.S. armed forces in the midst of transition. The Department of Defense had been established in 1947, and reorganized in 1949; the demobilization that had followed World War II had been traumatic for the services, and reduction and reorganization of the operating forces had followed drastic reductions in the defense budget. Tactics, weapons, and many of the aircraft used in Korea were of World War II vintage, and had yet to be integrated with jet aircraft, bringing about fundamental changes in operating procedures.

Props and jets of radically different performance capabilities flew together from the same carrier decks on the same combat missions. Helicopters also added another dimension to air warfare, and around the clock efforts to destroy enemy transportation resulted in a considerable amount of night flying by carrier fighter-bombers. Sustained interdiction and close air support combat operations by carrier aircraft over the Korean

Peninsula were a far cry from the war at sea experienced by so many of the navy veterans of the Pacific war who also fought in Korea. Instead of massive carrier task forces sweeping the seas in far-ranging operations against enemy islands or naval units, Task Force 77, a small force of two to four carriers, steamed with impunity in relatively fixed operating areas off the Korean coast. By the time the armistice was signed on 27 July 1952, U.S. Navy and Marine Corps aircraft had flown almost as many sorties, and had dropped more bombs and expended more rockets than they had in all theaters in World War II.

The Korean War was the first experience for American pilots in fighting a so-called limited war, in which there was a failure to press for a clear-cut victory and rules of engagement were imposed to prevent the war from spreading across the border into the People's Republic of China (PRC). Allied pilots engaged in aerial combat were denied the right of "hot pursuit" across the Yalu River into the PRC, and bombing runs against surface targets had to be conducted to avoid any flight across the Yalu. The situation was not only frustrating but hazardous for the flight crews, and, unfortunately, reflected a growing trend toward micromanagement of military operational matters by civilians in the Department of Defense and the White House.

Although the requirement to fight a war in the Western Pacific placed an extraordinary burden on the navy in terms of logistics, aircrew training, ships, and aircraft, there was no relaxation of treaty commitments nor suspension of skirmishes, crises, and tensions around the globe, most of which required some form of show of force or intervention by U.S. naval forces. The fifties and sixties were characterized by one crisis after another—Jordan (1956, 1957, 1963), the Suez (1956), Cuba (1956–58, 1962), Syria (1957), Berlin (1959, 1961), Nationalist China–PRC (1959), Guatemala–Nicaragua (1960), Dominican Republic (1961), Zanzibar (1961), Quemoy–Matsu (1962), Laos (1963)—on and on for round after round, decade after decade. If there was one common denominator that prevailed, it was the presence of naval forces, quite often centered around the aircraft carrier, that became the decisive factor in the resolution of a conflict or the easing of tensions.

The Cuban crisis of October 1962 has been called an icon of the superpower confrontation and a crucible for President John F. Kennedy. Perhaps no crisis brought the United States closer to a general war with the Soviet Union than this confrontation with its island neighbor. President Kennedy was quoted as saying that he had nightmares when he saw how close we came, and Soviet Premier Nikita Khrushchev remarked at a Kremlin reception that during the confrontation with the United States over Cuba, "We were close—very, very close—to a thermonuclear war."

The introduction of medium- and intermediate-range ballistic missiles capable of carrying nuclear warheads and jet bombers capable of carrying nuclear weapons into Cuba, allegedly in response to a perception that the United States was preparing to invade Cuba, presented the United States with a problem that was politically, if not militarily, unacceptable. The resultant U.S. naval blockade against ships delivering offensive weapons to Cuba and the subsequent return of Soviet offensive weapons to the Soviet Union led to a peaceful solution to the crisis, albeit with a Communist regime still in place just ninety miles from the Florida coast. U.S. naval forces operated in a highly professional and effective manner throughout the entire situation, despite increased meddling in the details of operational matters by the highest civilian authorities in the government.

A decade or so and many crises later, the United States again became involved in a war in the Western Pacific that could have been won but wasn't. U.S. combat on the Indochina Peninsula began in 1964 with the Gulf of Tonkin incident and came to its conclusion eleven years later. The war was in large part a naval and air war, with naval aviation involved from the very beginning to the final, hurried evacuation of Saigon in Operation Frequent Wind. The conflict precipitated a virtual revolution in air warfare; weapons systems, strategy, and tactics changed and evolved continuously as both sides refined the art of air warfare in the missile age. Electronic warfare systems and helicopters were among the weapons that had the greatest impact on the conduct of the war.

The burden of the navy's air war was carried by the aircraft of the U.S. Seventh Fleet, and a tremendous toll was taken of the aircrews, not just in terms of losses in combat, but in terms of the physical and mental debilitation caused by the high tempo of operations, family separations, and frustrations associated with fighting a war with their hands tied.

The overcentralization of civilian control reached its zenith in the Vietnam War, where the day-to-day conduct of the war was entrusted to inexperienced officials far removed from the conflict. Rules of engagement were imposed that hamstrung operational commanders to an unprecedented degree. Because of timidity and fear of conflict with the PRC or the Soviet Union, commanders and aircrews were placed in a precarious, no-win situation. Finally, like some chess game in which aircrews had been used as pawns by the "whiz kids" in Washington and the U.S. Military Assistance Command in Saigon, the war proceeded hesitatingly and indecisively toward an inevitable conclusion—defeat and withdrawal of U.S. forces from Indochina.

A navy A-4 carrier pilot summarized the war simply and succinctly: "The frustration comes on all sides. We fly a limited aircraft, drop limited ordnance, on rare targets in a severely limited amount of time. Worst of all we do all this in a limited and highly unpopular war."

The events of 1975 seem like a microcosm of the previous three decades. On 29 April 1975, in a period of three hours, U.S. Navy and Marine helicopters evacuated American citizens from Saigon, under heavy attack from invading North Vietnamese forces, while carrier aircraft provided fighter cover. In May the USS *Coral Sea* (CVA-43) participated with other U.S. forces in the recovery of the American merchant ship SS *Mayaguez* and her thirty-nine crewmen, illegally seized by a Cambodian gunboat. In October the USS *Inchon* (LPH-12) and five surface vessels served as a contingency evacuation force, with the USS *Kennedy* (CV-67) in support, as U.S. citizens were advised to evacuate their dependents from Lebanon because of political instability in the country. With the end of hostilities in Southeast Asia, the focus of U.S. military involvement would shift to the Indian Ocean and the Middle East and the pattern of crises response which began immediately after World War II would continue on unabated.

In June 1950 Lt. Gerry Miller joined the staff of commander, Carrier Division 1 as flag secretary and aide for Rear Adm. Eddie Ewen, who, as commander, Task Force 77, would be responsible for carrier air operations against North Korea. Inspired by the superb leadership of Admiral Ewen during this first of two combat tours to the Western Pacific, Miller absorbed Ewen's philosophy for living—"I'm easily satisfied with the best of everything." Vice Admiral Miller's thoughts on the Korean War are given from two widely different perspectives—as a staff officer privy to the command decisions that determined the daily course of combat operations at sea, and as the commanding officer of a jet fighter squadron operating from the USS *Princeton* (CVA-37). His description of nightfighter tactics in an F6F Hellcat provides a rare picture of this phase of carrier flying that had changed little since World War II, but would undergo radical transformation in the next decade.

Capt. Jimmy Thach brought years of combat experience in World War II to the bridge of the USS *Sicily* (CVE-118) as he commanded the ship with its contingent of marine Corsairs during fourteen months of operations in the Korean War. Thach paints a vivid picture of marine heroics as they flew sortie after sortie in the worst possible weather, providing close air support of ground troops. Though some of the tales seem apocryphal in nature, they make for enjoyable reading and bear witness to the well-deserved reputation of marine pilots as the best practitioners of close air support in the world.

Lt. Jerry O'Rourke was the leader of a five-plane VC-4 detachment of F3D Skyknights on board the USS *Lake Champlain* (CVA-39) during the

Korean War. To his alarm and dismay, he and his small band of aircrews and enlisted personnel found themselves unappreciated, unwanted, and in search of a mission. Undaunted, they traded the civilized life at sea for life ashore with marine nightfighters at a forward airbase, K-6, in South Korea. There, they found primitive living conditions and an uncertain future, but a warm welcome, respect, and camaraderie totally absent on board ship. The marines not only understood what nightfighting was all about, but were the leading practitioners of the esoteric art and science of aerial combat at night in the early years of the jet age. Jerry O'Rourke and his crews applied the lessons learned in the friendly skies over southern New Jersey to the deadly games of cat and mouse with North Korean MiGs while escorting U.S. Air Force B-29s on their nightly raids over the north. O'Rourke unabashedly recounts his love affair with the F3D Skyknight, while downplaying his own considerable accomplishments as an innovative nightfighter tactician, and proclaiming his undying support and admiration of marines everywhere.

At least some of the lessons learned in the Bay of Pigs debacle in April 1961 were applied to the Cuban crisis that followed in 1962. According to Adm. George W. Anderson, then chief of naval operations, contingency planning for the Cuban crisis was professional in approach and execution, various strategic and tactical options were considered by the Joint Chiefs of Staff, in consultation with the secretary of defense, and there was a good flow of intelligence and information to the operational commanders. In this regard, however, the feeling persisted that there was an inadequate flow of information to the Joint Chiefs on the behind-the-scenes sensitive negotiations between the White House and the Soviet Union. Throughout the entire process, Secretary of Defense Robert McNamara demonstrated an unusual preoccupation with detail, constantly meddling in the operational routine of the navy, and forcing Admiral Anderson to go to great lengths to prevent interference by McNamara and his staff with naval operating forces.

By the time Vice Adm. John J. Hyland assumed the job of commander, U.S. Seventh Fleet in 1965, the Cuban crisis was but a memory, and the United States had become inexorably involved in a land war in Southeast Asia. During his two years on the scene, Admiral Hyland directed the efforts of U.S. naval forces during an expansion in scope and severity that was without precedent in modern naval history. From his perspective as Seventh Fleet and later Pacific Fleet commander, Admiral Hyland discusses his deep involvement in the war during a five-year period in which U.S. combat operations reached their peak intensity. He discusses the navy's participation in the war, command relationships with other senior officers in the navy and the other services, the rules of engagement that were the source of so much bitterness among the flight crews, and the operational routine of the carrier task force on Yankee Station. As he remembers these final years of his distinguished naval career, Admiral Hyland's thoughts reflect the feelings of other naval commanders: "If I could do it again, I'd do it a lot better. But you only go around once."

What were the reactions of the senior officers on board the *Enterprise* (CVAN-65) to a strange message from the *Pueblo* (GER-2) calling for help? The commanding officer of the *Enterprise*, Capt. Kent Lee, former air wing commander during the Cuban crisis, had never heard of the ship, and had no idea where she was, nor had the task group commander on board, Rear Adm. Horace H. Epes, Jr. Admiral Lee tells the story from the operator's viewpoint, and also tells of a humorous, yet frightening, example of direct control over operational units from Washington when he receives sailing orders for his ship from President Lyndon B. Johnson by voice radio. The *Pueblo* story is but a footnote to Admiral Lee's compassionate, poignant insight into the daily lives of ship's company and the aircrews who learned how to deal with grief and maintain their equilibrium and composure during endless months of combat on Yankee Station. He speaks for all those who ever made back-to-back deployments to Southeast Asia when he laments, "It seemed as though it went on forever."

The story of Adm. William Lawrence's distinguished naval career is one of outstanding achievement in every field of endeavor, from athletics and scholarship at the U.S. Naval Academy to operational and staff commands at the highest levels in the navy. His view of the air war in Southeast Asia was from the front seat of an F-4 Phantom—a world of SAMs, AAA, and night carrier operations on Yankee Station—until a fateful day in 1967 when his aircraft was hit by enemy antiaircraft fire during a bombing mission over North Vietnam. Forced to eject over enemy territory, he was captured and spent six years of brutal, degrading treatment in a North Vietnamese prison camp. His chronicle of those years is a story of patriotism, personal bravery, and inspirational leadership. In Admiral Lawrence's powerful story of prison life, names of places, people, and events evoke memories of a traumatic era in American history: Hanoi Hilton, Haiphong Harbor, Jane Fonda, Richard Nixon, Jim Stockdale, John McCain, Jerry Denton, the Son Tay raid, and so on. Throughout it all, Bill Lawrence is sustained by his faith, competitive instincts, and resourcefulness, and he emerges from the ordeal with honor and dignity.

13
Korea
The Carrier War

Vice Admiral Gerald E. Miller

When Korea started in June 1950 we were in poor shape as we tried to mobilize somebody to send to the Western Pacific. We were issuing verbal orders to officers to report to aircraft carriers in San Diego. We'd just call them up, tell them to go, and send them a set of orders later. We didn't really have a very well-structured plan. One aircraft carrier was out there in the Western Pacific; that was it.

The twenty-fifth of June was a crucial date, as I recall. I remember the date because I had taken considerable action to ensure that when I graduated from Stanford in June 1950 I would be assigned to a nightfighter squadron. I had achieved that objective and had already reported to the squadron as executive officer. I was checked out in the airplane and tentatively lined up to lead a nightfighter team into the Pacific on one of the carriers sometime in the fall of 1950. It was obvious that night and all-weather flying was the way that we would want to be able to fight a good, full-scale war at sea. If you couldn't operate an airplane at night in all weather conditions, you were going to be severely restricted. But it was going to be pretty dangerous. A lot of us at that time were pushing for that program, and we had a good squadron at Moffett Field.

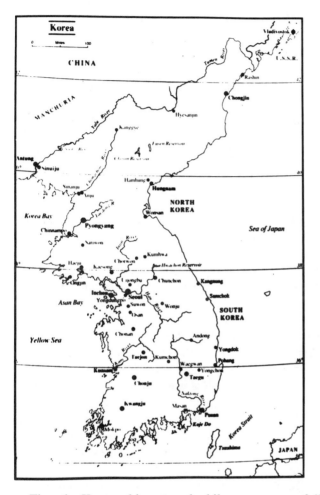

Map of Korea

Then the Korean thing started while we were woefully short of people. The navy was way down in size, below three hundred thousand men. Our carrier division staffs (admirals and people to coordinate the carriers) were down to just a couple. When the authorities decided we had to activate another carrier division staff in a hurry, I was one of the first to get a set of orders to move. I was to commission a new carrier division staff, and a wonderful guy named Eddie Ewen, class of 1921 at the Naval Academy, was the admiral selected as the flag officer. I was ordered down to be his flag secretary, to go to San Diego and get his staff commissioned. I then started a relationship that I consider to be one of the greatest I've ever had with any human being—Admiral Ewen. He came to San Diego and we began to get a staff together. By about the end of the first week in July, we had the frame-work. Then it was decided that Ewen should go out to the Western Pacific immediately and take over as the commander, Task Force 77. He would take over the staff that was out there already. Some other admiral would come in and take over the staff that we had started to put together in San Diego. Chief of Naval Operations Adm. Forrest Sherman wanted Ewen out there to take charge, and you could understand why when you reviewed the personality of the admirals who were involved in the situation. Ewen was a superb leader, absolutely superb; during World War II, he had commanded the carrier *Independence* (CVL-22), which was restricted to night operations.

He and I had about thirty-six hours together in San Diego, in getting a staff together. That's all. I knew about him, but we just had a thirty-six-hour association. I was properly responsive to what he wanted done. I didn't bug him about things, and all of the good lessons of the past evidently came through, because when he shoved off, he said, "You pack your bags and fol-low me out. I'm going to take you with me." So I left the rest of that staff, and followed him in about two weeks. It was around the middle of August when I showed up in Task Force 77. He'd been there a couple of weeks when I got there and moved in to be his flag secretary and aide on that staff.

I want to talk about this man. He had a great philosophy for living. He never had any children. He had a wonderful wife, whom he had known while at the Naval Academy—Betty Ewen—a beautiful, wonderful person. Their association was great to observe. She was a loyal supporter, followed him everywhere. She got as far as Honolulu on the Korean thing and could not get any further. She was always sending letters with all the clippings, files, and everything that was going on in the United States. She was his source of outside knowledge about what was happening back in the United States, and he would share all this with me. We became great friends. His philosophies were expressed in simple little things, and I still find myself re-peating them and passing them on. A simple platitude about being satisfied. He would say, "I'm easily satisfied—with the best of everything."

Stop and think about that! That is a great philosophy. It didn't mean that he was going to scream and shout and wanted everything for himself. But he al-ways set the highest possible standard as the initial objective. That's what we always went for—the best. I don't care whether we were trying to run a strike over the beach, whether we were trying to execute a joinup of two big forces at sea, or just planning a dinner party or cocktail party. Whatever it was, we always were going to go for first-class standards. That would be the objective.

Now, if we didn't get to that standard, we remained satisfied with what we got. As long as we had worked as hard as we could to get the best, we stayed satisfied with the result—no griping. It was an interesting philosophy, and I have never failed to remember it. He had another that I have passed on, and that I used in lectures when I talked to young men. I've heard it many times since, but the first time I ever heard it was from him. He used to say, "If you're not having any fun, you're not doing it right." That applied to the job out there. "If we're not having any fun in the way we're waging this operation, we're not doing it right." We'd better look at how we're doing it—not gripe about the system, not gripe about somebody else. Let's look at the way we're doing it and see if maybe we haven't missed the boat somewhere.

He ran Task Force 77 for six or seven months until a more senior admiral came over after the Hungnam evacuation and all that action at the end of 1950. Then other admirals arrived. We finished our tour and came back to the States in the spring of 1951. That was a great association, and I value my experiences with him tremendously.

We started Korean operations with aircraft carriers as we had finished in World War II. We would go in for a hit-and-run raid and then retreat as far as we could. We'd think about it and then go back in again. But it became obvious very rapidly that a transition was needed; sustained operations, day after day on the line, was the way we would have to run the Korean War. So we started operating in the Yellow Sea, staying there in a continuous operation. We had all the hassles with the air force about getting weapons on the target, because they had to come from Japan.

We had the agony of the Inchon invasion, whether or not the ships would be able to make it into the harbor because of the high tides. Ewen was a strong supporter of that operation. Adm. Arthur D. Struble was very good and had his neck out a mile. That was really one of his areas of expertise; he was a fine amphibious sailor, knew it well, and had the courage to go through with the thing. Of course, it was one of Douglas MacArthur's great ideas, and it was obviously the right thing to do. It was a fascinating period. Ewen ran that carrier operation well in support of Inchon. Then when we came to where we thought we had the Korean thing finished, the admiral and I and two other staff officers went off to Tokyo. After a couple of days, we piled into an airplane and came back to Korea to view the victory ground. We landed in Wonsan, among other places.

I remember landing in Wonsan in a special P2V aircraft that was configured for the top brass when they came out to view the battlefield. Then, the Chinese came down from the North and suddenly we were back aboard the carrier, back in the fight in earnest. The Hungnam evacuation, retreat, withdrawal—whatever you want to call it—when they called the marines down from the Chosin reservoir. We were involved in that deeply. We watched that operation day after day. The thing that Ewen did for me personally during the entire time we were on the line was trust me to run the task force. He would be on the bridge starting at four o'clock in the morning until about eight o'clock at night. He'd be up on the flag bridge watching operations, just being on hand. The chief of staff came on about eight or nine o'clock in the morning. He would be around throughout the day and would stay on the bridge until eleven-thirty or quarter of twelve at night. That left the period between midnight and four o'clock without the admiral or the chief of staff on the bridge.

I was, at that time, thirty years of age. That admiral and chief of staff trusted me. The rest of the staff was working hard on operations. I was just the flag secretary; I just had the action reports and routine administration to take care of. From my World War II experience, I knew the bridge routine of ships, so I ended up being the watch commander. They asked me to take the mid-watch. So for about the first six or seven months of the Korean War, when we were on the line, I stood the mid-watch every night from midnight until four o'clock. The admiral and chief of staff were both in their bunks during that time.

There wasn't that much activity going on during those hours, other than night steaming. We had course changes to make. Periodically we would run from the Sea of Japan to the Yellow Sea or vice versa, through the Tsushima Strait, and we would do it at high speed. Quite frequently we would run at 25 or 27 knots through the strait at night. I would be calling the courses, giving the signals, and I know that many people on those other ships would have been very concerned had they known that a thirty-one-year-old lieutenant commander was there—the guy with his eyes on the radar scopes—running the whole thing. But those gentlemen trusted me, and that's where I learned about "special trust and confidence," which is in the commission, a basic part of the profession. Of course, they knew I would be sensible enough to call them and get them out on the bridge if anything unusual came up.

It's amazing how things of that nature translate into the future. Later on when I was a carrier division commander in Task Force 77, during the Vietnam flap, one of the most exciting things that I had to conduct was an exercise to demonstrate our support of the Koreans. We operated in the Yellow Sea and the Sea of Japan. Here I was now an admiral on a big aircraft carrier, the *Ranger* (CVA-61), with many ships around. We were trying to probe the Soviet and Chinese intelligence systems to see what we could get away with in the way of task force operations. Here I was an admiral running through the same strait with a big ship, at 27 knots, just as I had when I was twenty years younger, back in the days of Korea. I never forgot what those gentlemen gave me in the way of lessons in trust and confidence.

So I spent the first year of the Korean War with that carrier task force. We put together a monstrous report, which is in the archives. That was an interesting period, when many changes were taking place. We made lots of mistakes; we were introducing jets into aviation at that time while we still had the straight deck carrier. We had all kinds of accidents and difficulties in using the airplanes with straight decks at that time—lots of miserable crashes, fires on the front ends of those carriers with planes going through the barriers. We really needed the angled deck badly at that stage and didn't have it.

We did have constraints put upon us about going over the Yalu River. Pilots were not allowed to go over the Yalu, even in hot pursuit, which was very frustrating. A MiG could come down and make a pass at you. You could turn around and get on his tail, and he would start to run for home. You could chase him up to the Yalu, but were not allowed to cross the river.

I've watched a couple of squadron commanders stand on the bridge in front of the admiral with tears in their eyes, saying "Admiral, I went across the river. I didn't know I was going across the river. I was bent on pursuit of this guy." And the admiral would say, "Okay, forget it." This was because of fear that the Chinese would come in. It was all part of the desire to keep the war limited. When the pilots were attacking the bridges crossing the Yalu, they could not cross into China and dive back across the length of the bridge into North Korea. They had to go down the river and dive at the bridge in the worst possible way—across, not down the length of the bridge. It was frustrating. Of course, the Chinese quickly learned about it, so they placed their antiaircraft batteries on the north side of the Yalu and fired at will, knowing that we weren't going to go after them.

That was the start of a new philosophy in how to fight a war. We could attack anything in North Korea. We never had any restrictions on that area, but we could not cross the Yalu; that was forbidden. The United Nations had a contingent in Japan at the time, in Tokyo. They had some kind of representation in Seoul, as I recall, but they never really got into the military aspect of the war. The reason I know they had a contingent in Japan was because we would visit there periodically. We'd get a five-day R&R period there every couple of months or so in Tokyo. We'd go to the high offices and meet some of the people. This was when Arleigh Burke was there with Turner Joy. Burke was a young admiral at that time counseling and keeping things straight. Eddie Ewen and Arleigh Burke were great friends. We'd get involved in a bridge game with him now and then. They were a great experience to be with, those two gentlemen. They were superb people.

I did get the experience of the Hungnam evacuation, which was a great evolution—very successful, moving all the marines plus a couple of hundred thousand civilians as I recall. We took them off that beach under fire from the Chinese. The civilians were relocated from North Korea to South

Korea. It was a superb operation. The mine field clearance in Wonsan: Struble ran that from a battleship in Wonsan Harbor. That was another fascinating operation. That was a very tricky thing; they did it all with helicopters. During the earlier attempt at an amphibious landing in Wonsan, the reports will show that we tried to clear a channel through to Wonsan by using conventional ammunition, dropping bombs from carrier aircraft. We felt that if we could get good navigational fixes, fly the airplanes down a straight line to the beach, dropping 500-pound bombs in the water, set to go off under water a bit, we'd blow up the mines and clear the channel. We got a couple of air groups out to do this, but it never was effective. We never could be certain.

Admiral Struble supervised the clearing of that one personally. He got all the helicopters we had in the fleet, and put them on the stern of that battleship and operated in Wonsan Harbor. By the time the field was cleared, the need to put the marines ashore had passed. Troops had come overland into Wonsan and we didn't really need to get in there from the sea, but we cleared the mines anyway. They were moored on the bottom in the harbor, and you could see them. You'd fly over them in a helicopter, and you could see them under water as long as there weren't too many waves. They were all Soviet mines. That mine warfare story in Korea is pretty fascinating; that was one of the biggest mine fields that has ever been laid, and it was effective—very difficult to clear.

We had some marines operating off of one of the aircraft carriers, the *Sicily* (CVE-118), with Capt. Jimmy Thach. We had them with Task Force 77 for awhile, but then they were put ashore. Once they got an anchor there, they worked directly with the ground division most of the time. Later, when Vice Adm. J. J. "Jocko" Clark arrived as commander, Seventh Fleet, the "bomb lines" were more firmly established and marine air worked for the Fifth Air Force—like most of the rest of us.

So the first year of Korea, actually the first eight or nine months, I was there with that staff with Admiral Ewen. When we came back to the States, he was given the job in Alameda as head of the Fleet Air Alameda program and he asked me if I would come up there and be on the staff for a period of time. So I went up to Alameda with him, worked about six or seven months as his personnel officer on the staff; then I was given the opportunity to get command of a jet fighter squadron and go to Korea.

In early 1952 we did not have jets in the nightfighter program. We were still flying propeller airplanes—F6F-5Ns at that time. They had a small radar, with a capability to detect a target up to about five miles. You relied on surface air control radars and air control procedures to get you within detection range of the target. The training program in Hawaii, designed to teach you how to get to the proper position and then close for a kill, ran about a hundred forty flying hours and usually took three months to finish. I had asked to get an accelerated program because I was going to be the commanding officer of a squadron. Not many squadron commanders took the course. I worked this out, and thought it would be a good background.

When I finished the nightfighter training program, I was really well squared away in propeller aviation. But I had flown only one hour of jet time when I became the commanding officer of a jet fighter squadron. Now, this was not unusual at that time. I had talked a squadron commander into letting me fly one of the jet airplanes in his squadron. I took some training films home and studied what makes a jet airplane fly, what makes a jet engine go; if you just get yourself a movie projector and draw all the navy training films, you can learn to do anything. I studied all the movies, got a checkout and flew one hour. Then I went off and did the nightfighter course. When I came back, I took over a squadron that had just come back from combat in Korea. It had some bright young fellows in it, wonderful young men who were now combat qualified. Here I was the commanding officer and had never had combat in aviation. Once again, I was learning from the junior officers.

I was assigned command of VF-153 and had the great fun of getting it ready for combat. Of course, I emphasized all-weather flying and night flying a great deal, although it was a daytime aircraft. But it had enough instruments in it to do a credible job, and that carried with it many rewarding experiences. We deployed to combat in January 1953 on the *Princeton* with Air Group 15, and that was the beginning of the last phase of the Korean War. We had thirty-two of the F9F-5 Panther jet aircraft in the group. It was the same basic airplane we had in Korea in 1950, except it had a bigger engine, so we got a little more thrust out of it. But many times we'd start with 400 pounds of bombs. Before we got off the catapult, they decided we'd better unload a couple and we'd launch with maybe 200 pounds plus a full load of ammunition because it just didn't have the thrust or capability to get off the carrier unless you had a lot of wind. There were many frustrations, but it was good flying. Many days we would fly two hops over the beach plus one combat air patrol—very rewarding from a flying point of view.

The need for the angled deck became evident. In qualifying this particular air group for combat deployment, we had to complete the usual carrier qualification work. Normally we would have had a couple of periods at sea to work this in, but the first period for our air group got canceled for some reason. So we ended up with about two weeks before we were scheduled to deploy, and the air group had not been aboard a carrier. We hadn't qualified

Lt. Comdr. Gerald E. Miller, commanding officer, Fighter Squadron 153, on board the USS Princeton *off Korea, 1953.* (Courtesy Vice Adm. Gerald E. Miller, USN [Ret.])

in carrier landings yet. Everything else—all the gunnery, bombing, navigation, and night work—was finished but we were not qualified to go aboard a carrier. Some of the pilots had made a lot of landings in the past. Many had been in combat, but they had to get refreshers. Many had been six or eight months without a carrier landing, and many of us had never made a carrier landing in jets; that included the air group commander and all the squadron commanders.

We had two jet squadrons, sixteen airplanes in each. We were in a squeeze; we had to go to the ship and qualify. We got into one of those situations off San Diego that is disastrous. The wind is from one direction and the swell is from another. You end up with the ship having a real pitch as it is kept into the wind. If you took it out of the wind, then you had a crosswind component that was bad, particularly on a straight deck carrier. We had these two sixteen-plane squadrons, all these jet pilots to get qualified. We got the propeller squadrons well squared away quite rapidly. There weren't any

The USS Princeton, *with Panthers, Corsairs, and Skyraiders on board, steams with Task Force 77 off Korea in 1953.* (Courtesy Vice Adm. Gerald E. Miller, USN [Ret.])

great difficulties there, but jets on a straight deck in that situation were tough. We didn't hurt anybody, but we broke up a lot of airplanes.

Vice Adm. Harold M. "Beauty" Martin was the head of the Naval Air Force, Pacific Fleet at that time. He was very unhappy with the captain of the ship and the air group commander, because it was a pretty expensive operation. In those days an F9F-4 probably cost about two hundred thousand dollars, which we thought was a lot when compared to World War II fighters at about forty thousand dollars each. But here we were ready to go to combat and our planes were in considerable disrepair. They had to bring in a whole new suite of aircraft. I bring this up to highlight again the great difficulties you can have playing around with a carrier deck that is basically too small to handle the elements of the sea. A plain rough sea state can give them fits. We had that lesson burned into us during that one qualification period.

Once we got over that hurdle and got the new airplanes, we were off and into combat. This was under Admiral Clark, who was the Seventh Fleet commander. Everything was fairly routine. It was the third year of the war and good targets were few and far between. In the latter phases, the bombline activity got to be pretty active as the North Koreans were trying to get some negotiations going. The North Koreans started to come out on the roads a lot more in the daytime. Railroad cars started showing up, trucks were on the highways in the daytime, and we had some fairly good targets. They had to get the supplies to the front, and they couldn't wait until night to move them. They were trying to reinforce their troops and make some movement on the bombline so that they would have a better negotiating position. So they were on the roads a lot more in the daytime than they had been.

When it became obvious that there was going to be some kind of peace, there was this great pressure to get the bombline in the most advantageous position so the negotiators at the table could have a position of strength. The only way we could keep our advantage was to keep them from getting their material to the front. About the only way we could do that was with air power because we couldn't make any penetrations with ground forces. We drew the bombline and said we will not go forward of that into North Korea. So we couldn't even send troops on deep probes; we had to rely on air power. That meant once again the night pilots had wonderful opportunities. This was about May–June–July 1953. But until then, it was scarce picking.

Occasionally we would have a raid on a hand grenade factory or something like that in North Korea. Those were great events. We would go in with all kinds of support—"flak suppression," we called it. That was about the best role for the jets. They really didn't carry enough bombs to do much damage, but they did have 20-millimeter guns. They made a lot of noise, and you could make three runs while the attack aircraft were making their one run with their bombs. So we were chasing up and down the skies, doing yo-yo stuff. If we got out without anybody getting hit, it was a good raid. But again, it was a period of searching, trying to find targets. We'd sit up there, arc around for awhile, taking a look at things. Then we'd go lower and take a look. We'd come down with a stream of four or eight aircraft and couldn't see anything. All of a sudden, "Bang!" the shells would start flying around. They'd shoot at the lead and hit the number two man. My wingman was hit several times when they were shooting at me. I was very fortunate in that combat period. I got hit three or four times, but never very seriously.

This was the time when the North Koreans had really dug into the mountainsides. They had truck parks in caves in the hills. They were in deep ravines and hard to hit. That's where they sat all day long and at night they would come out. We were pretty much a day operating navy at that time; very few people were flying at night. We just didn't have the capability. We could feel it more and more as the activity increased.

We dropped leaflets in August 1953 over the villages, telling the conditions of the peace, that a cease-fire was going to be signed, and so forth. I remember one day we dropped leaflets in the morning and bombs in the afternoon. It was interesting that the pilots and leaders adjusted readily to the political situation. As hostilities drew to a close, we were most concerned about the survival of everybody. We lost about 10 percent of the pilots, in all air groups. We knew when we deployed that about 10 percent would not come back.

The ground defenses were very good. When you got close to the ground, you were going to draw fire. Nobody was going to knock you out of the sky if you were high, but you couldn't be very effective if you stayed high either. You'd go after a target not always sure there was a target, but you took a look. Before we would attack an assigned target, I'd go down with a wingman and make a low sweep to verify there was something there worth hitting.

I was involved in both phases in the war, one when Struble was in command, a non-aviator, and then with Jocko Clark, an aviator. There was a noticeable difference. Struble turned the carrier operations over to the Task Force 77 commander, particularly after his exchange with Admiral Ewen early in the Korean flap. Jocko Clark, having his background in World War II and being an aviator, was much more interested in controlling the air action. He had a greater input. There were the "Cherokee Strikes," which were named after him. They were one of his ideas on the approach to the problem of stopping the logistic flow.

He was much involved in what we were doing. He paid a lot of attention to the carrier work, more than Struble did, but it was natural. The tenor of operations was different, too. When Struble was there, we had several amphibious operations—not just the Inchon operation. We had amphibious thrusts along both the Sea of Japan and the Yellow Sea coasts. Then there was the mine clearing operation in Wonsan Harbor. So the nature of the operation was quite a bit different when Struble was the fleet commander. By the time Clark arrived, it was principally air support of ground troops. We did it routinely, day after day, going in either deep to hit some logistic support target or working directly on the bombline, or maybe doing "on-call" strikes of various kinds.

In those latter stages of the war, when the negotiations were really under way and we were sitting across the table from each other, we were all very interested in getting peace. You want it over with, particularly after you've been out there for awhile. The deployment became a drag. Jocko Clark was telling us that we needed to put the pressure on, and there was a great emphasis on getting strikes and sorties out every day, and we flew a lot. I remember we had one superb day on the *Princeton* when everything clicked. The weather was good and we got started early. Our flying cycle was supposed to finish about four o'clock in the afternoon. Then we were asked if we could launch another strike, a jet sweep of some kind. Of the thirty-two jets aboard, we almost all were in an "up" status—I think we put something like twenty-eight in the air. The next day we received a message from CNO or CinCPacFlt or some place telling us that we had just broken the carrier sortie record for a single day—and we didn't even know it. We just flew, but we flew more sorties that day than any aircraft carrier had ever flown before in one day. Of course, one reason was because we'd been there for a period of time. The emphasis was on getting the sorties out for the bombline negotiations.

In the final phases of the war, when there was great pressure by the North Koreans and the Chinese to move that bombline to a more advantageous position, they launched a massive ground campaign. Things got tough, some of the ROK units were decimated. A battalion would disappear over night, until they got communications established the next day. Clark was adamant about all air units working together, about working with the Fifth Air Force. During the squeeze on the bombline, he sent a message that is in the annals some place. "Now is the time for naval aviators to come to the aid of the Eighth Army," or something of that nature. He went in to the commander of the Fifth Air Force as I recall, and told him that he would report to him for duty with his carrier aviators—that he would take orders from the Fifth Air Force. I don't know that those were his exact words, because I did not attend the meeting, but I surely felt the impact of what was said. His way of reporting to the Fifth Air Force commander was to take two squadron commanders off the carriers in the Sea of Japan and send them to Seoul, to report to headquarters there. Their task was to relay to Task Force 77 every day the attack missions that Task Force 77 was to perform, including the specific targets that were to be attacked.

I was one of the two squadron commanders that Clark picked. I had to leave my squadron and turn it over to my executive officer. A good friend of mine was also picked from our air group, and for three weeks we never left the compound of the Fifth Air Force headquarters in Seoul. Every day at one o'clock, we participated in the daily planning meeting, where a group planned the sorties for all of the air activity that was going to fly the next day. There was an officer representing the B-29 strikes; another represented marine air; there was a man for the night program for the B-26s of the air force; there was a man planning the fighter sweep, so he had contact with all the F-86 outfits; and I represented carrier air, which sometimes comprised three or four carriers.

The fighter sorties were planned by the people at sea, but I was picking the targets, with an intelligence structure to support me. I was picking the

targets and assigning the numbers of aircraft for each attack, and I was working in concert with these other people. Also at the table was a representative
of the British aircraft carrier, although I was given control over the sorties
for that ship, which was operating in the Yellow Sea. It may be of interest
that the British aircraft at that time had severe operating restrictions. As I recall, they could only stay in the air about an hour and fifteen minutes. They
couldn't go very far and they didn't have much in the way of IFF equipment,
so they couldn't be under positive control of the radar direction net. By the
third year of Korea, you needed a good capability to identify yourself and
help the radar operator on the ground locate you, and get you into the traffic
pattern, so that you could get to your target.

We handled the British carrier by assigning an isolated section on the
western end of the bombline. Everybody else stayed out of that area. We let
them run that any way they wanted—picking their targets, and so forth. This
was somewhat frustrating to the British. I remember the young flight commander in Seoul would give me a hard time every once in awhile, about letting them get into the war. I pointed out that if they could get some modern
equipment, we could be more aggressive in what we assigned them.

The British have come up with some of the best military concepts—their
air control procedures, their use of radars, their combat information centers,
the angled deck, the steam catapult, the landing mirror—all of those wonderful ideas came from the British, but they have never had the funding to
really implement them to the extent that they would like. And of course, they
are still some of the greatest people in the world. When Jocko's message
comes out saying "Now is the time for the Seventh Fleet to come to the aid
of the Eighth Army," that was like asking them to rally behind the king or
queen at all costs. I can remember that it was rather disappointing for them
when they found out that our idea of aiding the Eighth Army was somewhat
less glorious than theirs. They took the charge much more seriously, and
they were to be admired for that.

For three weeks we sat in Seoul and took our orders from the Fifth Air
Force commander as to how and what we would do with naval air. This was
all a part of Jocko's desire to give the best coordinated air support we could
to the Eighth Army. I have great respect for the way the air force ran the operation. We had no real difficulty with it, although the biggest problem the
navy had was in communications with the carriers from Seoul. Initially, we
could talk only about the targets scheduled for the day after tomorrow for the
navy. That was because of the communications delays; we were trying to do
everything by HF and keep it secure.

So the first thing I did when I arrived was to announce that I did not want to be the only guy sitting at the table talking about strikes for a different day than the rest of the planners. If we were going to coordinate, we had better address the same day at the same time. There was no difficulty. I just unclassified the targeting info a bit and got the information out, using the communications circuits the way they could be used. The navy started talking about the same day as the air force, marines, and so on.

In the last phase of Korea I was about thirty-three years of age. I was working with air force colonels, majors, and brigadier generals. We had army and marine ground people there identifying areas where they wanted support. It was a real joint operation; it was combat and it was good training, day and night for three weeks. Then they let us go back to our squadrons and finish up the war. We dropped the leaflets, the last few bombs, and had another week cruising around in the Western Pacific. Then we came home; Korea was over.

I lost five pilots out of twenty-one in that deployment, all of them over the beach. They were all shot down. We lost about 10 percent of our air group. Because I'd followed the Korean thing so closely, I was well aware of the casualty figures. I had been there early and watched the squadron commanders react to it. I was of the persuasion that I did not want a lot of extra pilots in the squadron. We used to talk about pilot-per-seat ratios. Many were pushing for two pilots for every seat, or one and one-half, or 1.3—the ratios varied back and forth. I tried to operate on the theory that the fewer pilots I had, the more proficient they would be. I could keep them very proficient while they were out there, and I wouldn't have as high a casualty list. Also, everybody would get more flying out of it; we wouldn't need as many people. So I went out with twenty-one pilots and sixteen airplanes. I had a lot of pressure to take more, but I was able to prevail. We had trained rather well, I thought. We helped the other jet squadron get squared away. We could do the job fine. But casualties in combat don't come strictly on the basis of performance and capability. There are many factors involved.

The first two weeks we were out there, I lost three pilots of the twenty-one. Not long thereafter, I lost two more; I was down to sixteen pilots and sixteen aircraft. Of course, we got replacement aircraft very rapidly, but getting replacement pilots was more difficult. There was a lot of pressure on me to take additional pilots, but I said I wanted to hold at sixteen. Rarely would all of the sixteen airplanes be functional at one time, and I'd meet my commitments. They allowed me to stay at sixteen pilots for sixteen aircraft.

The reaction that I had to the death of those five pilots might be worthy of recording. I've thought about it a great deal and wondered why a man reacts the way he does to the death of another. These were all relatively young men with the exception of one. We had brought into the air group a more senior officer—in fact, he was senior to me. He was on the air group commander's staff. I agreed to take him into our squadron to fly with us. He had never had any carrier experience before—a wonderful individual, a great tennis player of international fame, a big man. But he did not have any carrier experience; he'd been in seaplanes and what not. His wife had died of cancer and he had no family responsibilities, no dependents or anything. He wanted to get into the combat phase of aviation, and we agreed to give him a chance at it. We trained him, and he flew in my organization as one of our twenty-one pilots.

I had told the air group commander that his training was going to cost us an airplane, which it did. His was one of those we wrecked when we were doing our carrier qualifications. He never really quite had the feel for it. I kept him with me closely in the flying phase; at that time, he was probably thirty-five. But all of his flying had been in seaplanes, and this was a hell of a thing for the navy to try.

I was skeptical about it, but I kept him with me in my own flying division of four. He flew on my wing for quite awhile. He was a good man, and I was trying to get him qualified to move into a position of command. Frequently, as we were heading in to a target that was not going to be too tough, I would give him the lead. There might be eight, twelve, or sixteen planes in the formation. I'd give him the lead, and sit on his wing. He had the chance, he had to figure out where the target was and then lead us on the strike, get us down and back up, and then back to the carrier. But, because of his habit pattern of having a lot of time to work out a solution, as you do in a seaplane, it was very difficult to bring into jet aviation. You just didn't have the time in that jet aircraft; you had to act rapidly. It isn't that the man was not capable of quick thinking; he just had a habit pattern in the air from an entirely different environment. We didn't have flying time in training to really get a new habit pattern developed, and it was very frustrating to work with him.

When I went off to Seoul for the three weeks with the air force, I was very careful to discuss with my exec this special case. I cautioned the exec to be very careful; that any strike leads given to the man should only be pieces of cake. Well, they let him lead one. During the flight, he must have forgotten that when you're first off the flight deck, you have a full load of bombs and fuel; that if you dive the airplane toward the ground, it's going to sink a lot more in the pull-out than it will when fuel is low and the bombs have been expended. On this strike, they were just over the beach and found a good target. Bang! Down he went with his heavy load and in the pull-out, he mushed right into the ground. That was the end of him.

Well, I was sad. He was a good man, a good friend—but I guess what I had in my mind was that we almost asked for it. We set ourselves up for it; I didn't feel responsible. We had lost three younger pilots before that; I knew them well and I admired them. They were fine young men, but I never became too closely attached to them; maybe they were too young. I couldn't see them as career people; they were in the service, doing a job. For some reason I didn't have the same attachment for them that I did for some others. My reaction to their death was again one of sadness, and I was shocked at the adverse results in view of what I thought was a high state of training. But there was no great personal emotion.

Then came the fifth man. He was one of the finest individuals I ever knew anywhere. A big, tall, attractive guy—a Naval Academy graduate. He'd come out of the Academy right into flight training, right to Korea, got his combat time, and came back. He'd been a real source of strength when I was reforming the squadron, getting it ready to go back again. He was a water polo player at the Naval Academy, superb swimmer, all those things—attractive wife, a couple of kids, one brand-new baby. I had seven of this caliber who had something like four or five babies between them just before we deployed. In between tours, they had almost enough time to have babies. Actually, it wasn't quite enough time, so I left them all behind in the States to be present when the new babies were born. Then they flew out to Hawaii and joined us as we moved on to combat. I took a chance on that, and I'm glad I did, because this was one of the young men. He was there for the birth of his second child.

Well, he got hit over Wonsan in the daytime. I was on board ship at the time, listening to the radio transmissions of the episode. He ejected, his parachute opened, and I breathed a big sigh of relief. He landed in the water and a helicopter came out from Wonsan to pick him up. We had a rescue outfit in there at that time. But a strong wind was blowing. He landed in the water but couldn't collapse his chute, and he drowned. They got the body, but he had drowned. Of course, we had lost several pilots in Korea because of the parachute rigs of the time. Later, we changed that; we got quick release mechanisms, and so forth, so that you could get out of the chute in a hurry.

I had lost four other men in the squadron, and had taken it rather well. It bothered me, particularly in my professional pride as a commanding officer; you didn't want to lose anybody. You wanted to have the best record in everything. But when I lost this young man, I literally came unglued; I really was emotionally shaken. I still remember walking the flight deck most of the night, crying like a baby. I stayed outside; I couldn't go to my room. I couldn't go to the Ready Room or anything else. One man did it to me. And to this day, I start crying when I think about it. It's amazing, the varying affiliations you can have for the individuals for whom you are responsible. In some cases, it's a purely professional response. In others, there is an intense emotional involvement. This man was younger than I was by a lot, but I admired him so much. There were others in the squadron that probably would have triggered the same reaction from me. But having lost five pilots, I began to feel that I was going to carry the percentage load for the total air group. As the tour went on, the losses came from other squadrons. They all fit into the 10 percent rate for most groups.

I never felt the same in the surface navy in World War II, maybe because it takes longer for things to happen. Relationships don't seem to get quite as intense. But in aviation, somehow, I found that as the pressures increased, particularly when you became tired, you would become less rational. Some strange thoughts would come into your mind. After a few days ashore when you could step back and take a look at yourself, you would wonder how in hell you could come up with such ideas. I can understand how men in tough combat situations where they may be under stress for long periods of time can come unglued. I saw that to some degree in World War II with men who had been at sea for long periods of time. They would get frightened. The word would come over the system announcing that a Japanese task force was on the horizon; general quarters sounded and you were charging out to man your battle station. After a period of time, some of them would break down.

It was battle fatigue; you could see it. You would see men break out in a rash, open sores on their body, from just pure tension and stress. Maybe they weren't closely involved in combat. They might be filling watch duties or acting as a supply officer on the Task Force 77 staff, just working themselves into the sick bay with nothing more symptomatic than a nervous rash. But the pressure comes on and you really have to question judgments. A man will do things under tension that he will never do otherwise. We always tried to get as many breaks on the battle line as we could. But for many of those periods, we stayed for a minimum of thirty days, and often it was fifty or more. We would have a break for logistics, maybe a day off, or something like that every once in awhile. Your balance, your sense of reason, shifts considerably under the constant pressure. I never forgot that. I learned in Korea that you really can't trust your judgment when you are tired—under fatigue conditions.

Korea was a great experience from the point of view of showing us how to do things right and wrong. There were a lot of wrong things in that campaign. I'm still fascinated with the MacArthur-Truman story and the events that led up to it. I have a feeling that particular fracas was a key point in marking the real decline of the authority of the military. I really think it was the most significant event. It has been said that perhaps it was aggravated in part by MacArthur's recommendation that they use atomic weapons. In looking back on it, I can see such a recommendation being accepted more readily then than now, despite the fact that we have a far superior capability now. But the other side didn't have any real capability to retaliate then. There could have been some validity in their use. History could show that it might have been better to finish the thing off.

I thought MacArthur's arrogance and disdain for civilian authority were interesting to observe. In some ways, you might say, "He showed them. He stood up to Truman." But in the long run, he may have done more damage to the military structure than anyone can imagine. He may have undermined the authority that had been so firmly established in the military during World War II. The Office of the Secretary of Defense was in existence and was starting to expand. All that was needed was some lack of confidence in the senior military, particularly in a man of MacArthur's stature, and the stage was set for what we have inherited today—the constant concern about civilian authority and control over the military.

There are more and more constraints imposed every year. When I think of the authority I had as an admiral and compare it to the authority I witnessed when I worked for admirals in Korea and World War II, the difference is absolutely amazing. In watching Ewen and Struble operate during Korea, in observing the MacArthur incident, and then seeing Adm. Elmo Zumwalt as the head of our navy, it became so apparent that mixing politics and the military together in a profession is a disaster from the military point of view. I do not believe a man can be a politician and a military officer simultaneously. We had MacArthur as a superb military leader. When he started to play politics, he got into trouble. He may have gotten all the rest of us into trouble for many years to come. You saw it in modern times with Zumwalt, who was politically motivated for many years before he ever got to be the head of the navy. As the head of the navy, being politically oriented, he played politics a lot. Now that he is retired, you see it coming out constantly. He destroys some of his credibility as a military officer by exposing his lust for political power—at least in my opinion.

People wonder sometimes why retired officers don't write and don't speak more. I guess a lot of it has to do with laziness. But I think much of it has to do with the fact that we don't want to bring discredit on the profession, to undermine it in any way. The only motivations for writing and speaking are to support the existing system, and the active duty people are in pretty good shape to do that. They are current and better prepared to do it than most who are retired.

=14=
The Black Sheep in Korea

Admiral John S. Thach

John Smith Thach was born in Pine Bluff, Arkansas, on 19 April 1905. He was appointed to the U.S. Naval Academy in 1923, was graduated and commissioned ensign on 2 June 1927, and was designated naval aviator on 4 January 1930. He served in a wide variety of carrier-based, patrol, and experimental squadrons, and had numerous command tours ashore and afloat as he advanced to the rank of admiral, to date from March 1965. Admiral Thach was commander in chief, U.S. Naval Forces, Europe from 25 March 1965 until his retirement on 1 May 1967.

During World War II, Admiral Thach participated in twelve major engagements or campaigns, operating from seven different aircraft carriers. He was twice awarded the Navy Cross for extraordinary heroism in aerial combat with Japanese forces. He later went to the Fast Carrier Task Force in the Pacific as air operations officer and developed the system of blanketing enemy airfields with a continuous patrol of carrier-based fighters and planned and directed the navy's final offensive blows against the Japanese homeland. During the Korean War Admiral Thach commanded the carrier Sicily *(CVE-118), from which marine aircraft provided close air support during many crucial battles.*

U.S. Marine pilots from VMF-214 debrief Capt. John S. Thach, commanding officer of the USS Sicily (CVE-118), during combat operations against North Korea, 8 August 1950.

Admiral Thach's decorations include the Navy Cross with two Gold Stars, the Distinguished Service Medal, the Silver Star, and the Legion of Merit with two Gold Stars.

I took command of the escort carrier *Sicily* at San Diego in June 1950. A few days later I was still alongside the pier at the Naval Air Station, North Island, when at four o'clock in the morning, Korea time, on 25 June, the North Korean army came charging across the 38th parallel and invaded South Korea. We had very little logistic capability in the Pacific; our carriers were down to the point where only the *Valley Forge* (CV-45) was west of San Francisco. So President Harry S. Truman, after a hurried conference and a very plaintive and convincing plea from President Syngman Rhee of the Republic of Korea, authorized Gen. Douglas MacArthur to use air and naval forces to assist. First, he stepped up the supply of military assistance. Of course, this was an extreme emergency and deciding to give the South Koreans more ammunition wasn't going to have much effect.

MacArthur finally recommended that he commit just one regimental combat team to throw in with the few people who were already there to try to help the South Korean army hold the line. Some people felt maybe they could, but that just wasn't in the cards at all and just one regimental combat team was, in my opinion, a gross underestimate of the requirements by General MacArthur.

Not much was done as far as ships on the West Coast were concerned in the first day or so because the only action that was taken was to commit naval and air help. Also, President Truman directed the Seventh Fleet to go and neutralize Formosa, as far as any combat situation was concerned—in other words, prevent Formosa from attacking the Chinese mainland and prevent the Chinese on the mainland from attacking Formosa. The *Valley Forge*, one cruiser, and eight destroyers—that was the Seventh Fleet, and that was all the combat power we had west of San Francisco. In view of all the talk that there would never be another amphibious landing, requests to maintain an amphibious capability had also been denied. So we were in a very weakened condition.

My supply officer had been on the ball because before we even got sailing orders to move he had been reading the newspapers, too, and had started putting in dry provisions and all the things he thought we might need to have, which saved some time. The *Sicily* was part of Carrier Division 15; Rear Adm. R. W. Ruble was commander, Carrier Division 15, and he had two ships, the *Sicily* and the *Badoeng Strait* (CVE-116)—we called her the "Bingding." They planned to sail the *Sicily* without any escorts and, on top of that, to immobilize the hangar and flight decks; they had so many airplanes they wanted to put aboard, we wouldn't be able to fly and here we were going clear across the Pacific and a war had started. I think it was a decision made by the urgency of the situation to get the *Sicily* moving because we were the first ship to leave the West Coast.

We arrived at Pearl Harbor and I got a good briefing on the war situation, which didn't look good at all. Other members of the United Nations had promised some aid, but we were in a situation where the logistics problem was a hellish thing, since there wasn't much of a service force in the Pacific Fleet and a lot of it would have to come from the continental United States—maybe some from Pearl but even Pearl Harbor didn't have too much to spare. The problem was to match the logistical ability of the North Koreans, who had short distances to go, with the distances we were involved in. It was going to take a tremendous effort.

We got to Guam on 20 July 1950, offloaded my ASW airplanes, then headed for Yokosuka. I got to Yokosuka and went immediately to Tokyo to get my instructions and a briefing. In this briefing at commander, Naval Forces, Far East (ComNavFE) headquarters, I learned about the great difficulty they were having with air support of the troops. They just didn't have any air control on the ground. They would scream for close air support and the *Valley Forge* would send in eight or ten airplanes and they couldn't, wouldn't, use them. Communications were all fouled up and everything was a mess. So they would have to just pick out their own targets to be sure not to hit our own troops and try to do the best they could, which is a very unsatisfactory way to do it. In other words, they just had to turn to interdiction on their own or dump their ammunition.

I was directed on 31 July to go to Kobe, where the *Badoeng Strait* was, along with the supply vessels that were carrying marine Corsair spare parts and their squadron personnel. On 3 August we landed the marine squadron aboard and flew our first flight in close air support. The orders I had were the best orders anybody ever received in his life. They said, "Proceed to Korea via Kobe. Render all possible support to ground forces. Direct air support or interdiction at your discretion. Keep ComNavFE informed of actions, intentions, and position." This gave me complete freedom to go in where I could help anybody I wanted to. So that's what we did.

I had one squadron of twenty-four airplanes aboard—VMF-214, the Black Sheep squadron. That's about the number that you could operate most efficiently, where you had plenty of room to move them around, and that was the size of a marine close air support squadron. Later, another squadron, VMF-323, relieved them. They were also real pros, but that was much later in the Korean War.

These marine pilots of VMF-214 were all quite experienced; they weren't young kids. Most of them were married and had children, and they took their work seriously. They really were the top pros in the business, I think, in the whole world. Many of them had been in World War II. They were heavily decorated and they knew the business of close air support. Every one of those pilots had infantry training. From time to time later in the Korean War, pilots from the *Sicily* squadron would rotate into the front lines as air controllers for a while and then back to the *Sicily*. The marine troops knew them by their first names; they knew who they were and what they could do.

The close air support problem as it existed at the time had a bearing in a way on why things went the way they did. Of course, the U.S. Air Force wasn't too interested in close air support. The people in high command in the air force were primarily focused on the big bomber idea, that you really didn't even need troops to win a war—just fly over and bomb them, and then wait for a telegram saying that they surrender. So the air force was utterly unprepared to do close air support the way it had to be done if you were going to help the troops at the front lines.

It wasn't a matter of just being a communication problem; it was a matter of education over a long period and experience and doctrine built up. I just couldn't believe it could be so bad, and neither could Rear Adm. J. M. Hoskins, who commanded Task Force 77. He'd send these planes down and the pilots would come back and say, "We couldn't help. We wanted to. We were there and we couldn't get in communication with people." Those that did would get in touch with the Joint Communications Center and they either wouldn't have any target or by the time they got it it was old information.

The air force went on the principle that any aircraft committed had to go through a pretty high echelon of command before it did anything. They wanted to keep tight control in high places. Never would they let anybody in the trenches control one of their airplanes; they were going to do it themselves from higher up. Well, they had to get some sort of a working arrangement with the carrier-based aircraft who were used to being controlled from the ground in the front lines. They did it on every amphibious landing, and their tactical air control squadrons had a great deal of history and experience, and the navy-marine teams were real pros in this business.

Another thing that made it difficult was that the Fifth Air Force wouldn't accept a navy tactical control party. Help would be offered by navy air control and they didn't want to take it. They were more interested in getting control of all aircraft than they were in helping the army! When General MacArthur and Vice Adm. A. D. Struble, commander, Seventh Fleet, went to Formosa, Gen. George Stratemeyer, commander, Far East Air Force, quickly held a conference. The official record of it says that they deployed four generals and a colonel to face one captain and two commanders and two lieutenant commanders to make sure they controlled the decisions that were to be made. Nevertheless, something good could have come out of this, but it didn't. They wanted to be sure that they would get operational control of all aircraft in the Korean theater.

As a result of the conference, they put out a memorandum saying that first priority for carrier operations would be in close support, second was interdiction south of the 38th parallel, and the third priority was to strike Bomber Command targets beyond that line, if requested by the Bomber Command. Coordination for the attack south of the 38th parallel was to lie with the Fifth Air Force. Of course, you'd have to get permission from the Far East Air Force to attack any targets otherwise, which were supposedly bomber targets. They interpreted this as operational control and this disturbed the navy quite a bit. It had implications that they would be telling the carriers where to go—right rudder, left rudder, and so forth.

We couldn't stand that and Admiral Struble, since he was not even consulted before this memorandum was put out, didn't want to go along with it at all. He tried to get away from it but there was not much he could do—he didn't get help from anybody. So the Joint Operations Center (JOC) was formed by the air force at Taegu, and the navy fast carriers tried very hard to cooperate because of these urgent and emergency calls for help by the army. The army must have been giving the air force hell at that time, because the air force even started repeating their screams for help to the fast carriers. So the carriers would send everything that they asked for.

I have listened on the close air support radio many times. By this time there really wasn't any Fifth Air Force, Korea. The Fifth Air Force was back in Japan because, as somebody said, the best way to defeat an air force is to walk into the airfield and keep the airplanes from flying, and that's what the North Korean army did. The enemy got so close to Taegu they were within the landing circle, so that field had to be given up. By this time the field at Pohang, we called it Pohangdong, was being exchanged sometimes every night.

I know one time we sent some planes over there because the army had a patrol that really wanted help and our pilots were willing to work at it. I heard this on the radio and also the pilots informed us; the army controller was so enthused about the *Sicily* Corsairs, the marine pilots, that before releasing the planes after they'd expended their ammunition, they begged them to come back the next day. The words were sometimes "Please, please, come back tomorrow. We'll take that airfield back again. If you'll just come back tomorrow we can do it together." It would almost bring tears to your eyes to realize how much these army troops over there wanted some real good close air support. They hadn't ever had it before. One of them said, "We had close air support like I've never heard of before. This is something I didn't realize could happen."

The F-80s would come over from Fukuoka, Japan, and the front line was just near the end of their range. They'd call the controller and say, "Give me a target, give me a target. I've only got five minutes more. Got to go back." They would be asked, "What is your ordnance?" "I've got two 100-pound bombs. Hurry up." I heard that so many times and finally I heard, I think it was an air force controller, say, "Well, take your two little firecrackers and drop them up the road somewhere because I've got something coming in that has a load."

On 7 August the First Provisional Marine Brigade, which had been landed at Pusan and got in position under Brig. Gen. Edward A. Craig,

Ordnance crewmen on board the USS Sicily *make final checks on F4U Corsairs prior to launching for a strike on Korea, 16 November 1950.*

attacked westward from Masan toward Chinju. This was the first time that the escort carriers with the Corsairs and the marine ground forces got into action with everything there for coordination. It was a beautiful thing to listen to. I couldn't see it but I knew what was going on. It was just like going from confusing darkness into bright daylight. The coordination was perfect and everything clicked just the way it should. You should have seen those pilots when they came back. Each one would heave a big sigh of relief and say, "Now we're doing what we're supposed to do in the right way."

Unless we were working with the marines, the pilots still had to go through that Joint Communications Center. Working with the marines, we didn't have to bother with the JOC, because we had our own tactical air control group. The forward air controller, the ground controller, often an aviator, would sometimes be out in front of the front lines. Sometimes they'd be in a little jeep or a tank or just crawling along and dragging the communications equipment, in the bushes.

One time the controller said, "I want just one plane of the four to come down and make a dummy run. Don't drop anything. I'm going to coach you on to a piece of artillery that's giving us a lot of trouble. I'm very close to it but I can't do anything about it. It's just over a little knoll." He described the terrain, just where it was and so forth. So the leader came down in his Corsair and he was coached all the way down, the air controller practically flying the airplane for him.

Then the controller said, "Now, do you see it?"

And the pilot said, "Yes, I see it."

"Okay, then, go on back up and come down and put a 500-pound bomb on it. But be very careful." So the pilot came down and he released his bomb and hit it with a big explosion.

The controller said, "Right on the button, that's it. That's all. Don't need you anymore."

The pilot said, "Just a minute. While I was on the way down, on the right hand side of my gunsight, I saw a big tank. It was under a bush, but how about that target?"

And the controller said, "I told you I was close. Let it alone. That tank is me."

By 12 August, this marine brigade was nearing Changchonni and the First and Second Battalions were ambushed. The enemy counterattacked and gave them a bad time; they gained about twenty miles in a few days to try to straighten out the lines. This area around Masan and Chinju was a flat area and most dangerous. The enemy could move in fast. It was the road to Pusan, so everybody wondered whether we could hold it. That's why they put the marines in that particular spot. Well, this ambush had pinned down some of the people. They had little pockets, and they had a number wounded in a fight that was pretty far out in front. We had close air support aircraft working on them, and they said: "We're going to send an ambulance in and get those people out"—send an ambulance right out in front of their front lines. "We want escorts for that ambulance. We'd like to have four Corsairs."

The four Corsairs reported, and here comes this ambulance tearing down the road and the Corsairs weaving right across it, spraying the bushes and anything that might stop the field ambulance. The ambulance got in there and brought all those people out and never got hit. But the Corsairs were really working over the sides of the road to keep any snipers or anybody else from sticking his head out and shooting.

On the afternoon of the twelfth they got orders to pull back and the marines were a little disappointed because they thought they would go ahead. They'd gained twenty-six miles by then in four days, but the order to pull back was because the enemy had gotten across the Naktong River in another sector. The army had counterattacked but it hadn't done any good. So they pulled the marines back, took some of them and put them into the breach again, into the place where the toughest fighting was and the enemy was being successful, to stop it—and they stopped it.

I know that at one place where they crossed the Naktong River was on a little wooded knoll. The enemy had gotten some automatic weapons in there; there was a ridge shaped like a horseshoe, and the marines were all around near the top of this ridge. Some of them were right on top of it, looking across at this little knoll where the enemy was, but just a little bit out of rifle-fire range. They wanted to bring in a division of Corsairs to direct them and the air controller was right there on the ridge looking at the knoll, to bring these people down.

They had to fly right down over the ridge then start shooting right away, or start shooting really before they got to the ridge, with rockets. After the first one came down, the marines on the ridge were standing up and looking; they wanted to watch the rockets! The last pilot that came down after they'd put these rockets in there called the controller again and he said, "Would you please have the people in the front row be seated. I can see the back of their heads in my gunsight and it makes me nervous!" So he said, "I will tell them to sit down in front." That's how close these Corsairs were.

By 1 September 1950 the plan for Inchon was well under way, and on the tenth the *Sicily* and the "Bingding," under Admiral Ruble, were directed to go up off Inchon and burn the west half of the island of Wolmi-do. Wolmi-do was an island from which a causeway ran into Inchon, to the harbor. So we went up there and put napalm on the west half of Wolmi-do and burnt it up so it looked real naked.

On the eleventh we replenished, but on the thirteenth we were back in there, starting the softening-up process. Most people don't realize how much time is spent before an amphibious landing, or any kind of landing of troops, by the navy in softening the place up. This is really one of the secrets of a successful amphibious landing. So, near the end of August and the first part of September there was this softening-up process. Destroyers and cruisers went right in close. Finally on D-minus two or three they pushed cruiser 8-inch and destroyer 5-inch shells into Wolmi-do and other defenses, and finally a whole train of amphibious effort was under way.

D-day on the fifteenth came along and there were typical H-hour prelanding strikes. The marines landed on Wolmi-do and thirty minutes later had taken it completely. They breezed across the causeway there, with destroyers standing by with influence fuses, antipersonnel shells in case anybody tried to stop them. Then they moved into Inchon and took Inchon very quickly. The opposition stiffened when they got in through Inchon, but the marines were on their timetable and by some very magnificent work on the part of the Service Force supplies arrived on time. Some old decrepit LSTs and other World War II things that should have been scrapped a long time ago got in to the beach at Inchon. Some got hit by enemy fire, but were towed or pushed by other craft and they got in there. They didn't break down of their own accord.

This was a wonderful thing because it gave the marines enough ammunition to keep moving at the rate they wanted to move. On the road between Seoul and Inchon there were some Russian-built and -supplied tanks moving down, so we got an urgent call on those. The marine artillery was real good. They knocked off three out of the five and the *Sicily*'s Corsairs got the other two. That was D-plus one, the sixteenth.

Then there were some more tanks on D-plus two. As a matter of fact, D-plus two was a very busy day for the First Marines and all their supporting aircraft. Four more tanks were destroyed en route to Seoul. By that evening, though, the marines were within fifty and one hundred yards of a little place called Sosa, and the Fifth Marines had overrun Kimpo Airfield, which was the big prize of the day. That was what we wanted, that airfield. They occupied the high ground overlooking the field and had troops on the landing area itself by nightfall.

People wondered what was the best ammunition to use against the tanks. The *Sicily* had received the first of a recent development by the navy at the West Coast proving ground called a shaped charge that you could put in a shell or a 5-inch rocket. When the explosion occurred it all went forward, so it had an order of magnitude higher penetrating power than if that similar charge were not a shaped charge. These shape-charged rockets were used very effectively, although it was a little harder to hit a tank in the right place with them than it was to come down and lay a napalm tank on them. Actually, napalm, we discovered, was extremely effective against tanks because you could scoot this in right under the tank—even if it was moving you could get in close and hit it. There was no shrapnel explosion so you could come in very low, just skim over the top and let the thing go, and the crew of that tank just came out of it like popcorn. There are many things on a tank that burn; it has fuel and oil and so forth in it and when it gets hot enough the ammunition starts going off. So it almost blows itself up.

We were in operation a full thirty days without any rest or recreation. By 30 September the troops were almost back to the 38th parallel; they had retaken Seoul. We departed Inchon with the *Sicily* on 3 October and headed for the Sea of Japan with orders to operate in support of the Wonsan landings and the general advance into North Korea. We were engaged in this sort of support business until 17 November 1950. Then the decision was made to revert the *Sicily* to ASW training operations in the Western Pacific area. The reason for this was that General MacArthur was very encouraged about the progress of the advance and he was going to try to clean up the North Korean army. Things were going very well from his point of view, and for some time the navy had wanted to get these ASW carriers back in the ASW business.

So we went in to Wonsan to offload the Black Sheep squadron. This was a rather sad parting of two members of a team that had worked so beautifully together, the marine squadron and the crew of the *Sicily*. It was a very cold day when we anchored in the harbor, after running a narrow swept channel through the mine field. When the marines, mechanics, planes crews, and technicians had gotten in the boat I was looking down watching them and feeling pretty sad about putting them off in this cold, bleak terrain and I noticed several sailors taking off their own gloves and pitching them into the boat to the marines. So I took mine off and threw them in also. This sort of illustrated the close feeling between the marine squadron and the *Sicily* crew.

Well, we offloaded on the eighteenth and then that evening got under way, came out the channel again, and headed for Guam. We arrived in Guam, picked up the ASW squadron, turned right around, and came back to Yokosuka, arriving at Yokosuka on 1 December 1950. All this time I had been making plans to bring up the state of the art of ASW as far as hunter/killer groups in the Western Pacific were concerned, to try to get it back up to where it had been, at least. But all this detailed planning came to nothing because within a day or two after I arrived back at Yokosuka I got sudden orders to offload the ASW squadron and proceed to Hungnam at best possible speed and reembark the Black Sheep squadron.

So we got the planes offloaded as soon as possible, got under way, and continued at flank speed all the way down around Kyushu and up into the Sea of Japan to Hungnam. I picked out a nice spot, close enough so that the boats wouldn't have too far to go when they brought out all the equipment that the marine squadron would have. They were flying close support missions for the marines that were up in the Chosin Reservoir area and it was important that they never missed a beat.

So we loaded the equipment all night. Of course, we took the ground crew aboard in boats and one big marine gunnery sergeant climbed up the ladder to the hangar deck, got down on his knees and kissed the hangar deck. It wasn't just a little peck—he gave it a real good "smooch" and you knew he meant it. Apparently that's the way they felt about it.

The pilots stayed ashore with the airplanes with the last load of ammunition and gasoline to fly a strike the next day from the field, and they managed this without a break. I got under way the seventh of December, they made a strike, landed aboard, and then we started striking again. The marines up at the Chosin Reservoir faced what seemed like an impossible situation. Major Gen. Oliver P. Smith, who was the commander of the First Marine Division, put out an operation order that said, "We will advance through the enemy toward Hungnam." There were from five to seven Chinese Communist divisions against them, and it looked like it would be impossible to get anybody out of there. The air force commander who had responsibility for airlift flew into a little field up at Hagaru and offered to attempt to fly the marine division out. But General Smith had different ideas. He said, "No, we walked in here and we're going to walk out." They were flying out casualties and what surprised everybody was that the marines were flying in replacements. Here they were trapped up at the Chosin Reservoir and they were flying replacements in.

In the eyes of some people, they were just getting more people in trouble, but that wasn't so. They started moving and they still faced a lot of harsh treatment from all these hordes of Chinese. There were places on the ground down there where you could hardly step without stepping on a corpse. I've seen photographs of the fields just strewn thick with them. They weren't out of trouble yet. They still faced that long, slippery road that stretched from Hagaru to Hungnam. They had numerous ambushes and, of course, the Communists would blow up any bridges that they could get to. A lot of the ambushes were busted by Corsair support. One of their controllers kept calling the *Sicily* planes the "ambush busters." He'd say, "I want four ambush busters," and they'd go in and work over an ambush.

The air control and coordination just couldn't have been any better; it worked beautifully. Everybody that came in was given a target expeditiously. I don't think much ammunition was wasted in that operation when they came down that long trail to Hungnam and finally got there. This was 7 December and the weather was terrible. There was snow all over the ground and the wind was cold. Operating the *Sicily* at sea, it was so cold that the saltwater spray would freeze, and we got icicles hanging all over the ship, the bridge of the ship and on the flight deck; but the flight deck crew did a magnificent job in keeping the catapults and arresting gear free and keeping the deck from being too slippery with ice. They'd go out there with shovels and hoes and break it up and rake it off.

The thing about airplanes based ashore on a field was that they had to service and maintain them in that cold, rain, and sleet. The guns apparently weren't working. After the first strike, I went down to the Ready Room for the debrief by the intelligence officer and nobody was saying anything about what they did with their guns. They were just talking about how they employed their rockets and bombs, so I said, "What about your 20-millimeter guns?" They said, "Oh, we haven't been firing those for a long time. They're just not worth the effort and the risk of making a low pass when your guns aren't working." So I said, "Well, we'll fix that. We'll just take a few planes at a time, starting right now, and clean up those guns real good, and any amount of manpower help you people need you can have it." Inside of three days we had all the guns working again.

The cold weather had made it almost impossible to use their hands. Their fingers were so numb, and to take care of these guns, take them apart, clean them, and put them back together again, in that cold weather, they had an awful time. The first thing I noticed about the pilots was that they didn't smell very good, and that's understandable. Some of them said they hadn't had their shoes off for twenty days, and they looked hungry. They weren't in too good shape, but they still had terrific morale and they were going to keep those planes in the air if it killed them.

So the marine First Division advanced toward Hungnam and got there. On Christmas Day, 25 December 1950, the *Sicily* and the *Badoeng Strait* were able to depart the Hungnam area and we were sent around to the Yellow Sea. On 27 December we arrived off Inchon, of all places, again, and the *Sicily* relieved the British carrier *Theseus*. The *Badoeng Strait* went on down south to Sasebo, Japan. She had some repairs to be done, and it was the plan to rotate us anyway up there on the line.

We started to fly missions in support of the Eighth Army. The Eighth Army then was getting worse and worse into difficulty and by 4 January 1951 they had to abandon Seoul again and they blew the bridges over the Han River. I don't know how many times those bridges over the Han River have been blown. But the thing that made me feel so bad was that their close air support organization was just nothing. I had hoped there would be some improvement from the previous times that we had attempted to operate with them. There were some instances where you could get with the controller by great patience and do some good work, but it was just not organized like the navy-marine team and I think that the close air support—and a marine general said this, the one with the First Division—if the close support hadn't worked as well as it did they would never have gotten out of the Chosin Reservoir. They would all have been killed or captured by the Communists because it was just like having artillery right over your left shoulder.

On 7 January 1951 the *Theseus* came back and relieved the *Sicily* off Inchon. On 20 January we disembarked VMF-214 and on the twenty-third sailed for the United States for dry-docking in Long Beach and minor overhaul. On 5 February we arrived in San Diego en route to Long Beach, and in May the *Sicily* completed its dry-docking and overhaul. A number of things that had been troubling had been repaired and some new equipment was installed, then we returned to the Western Pacific.

On 2 June 1951 we were assigned a different squadron, VMF-323. The thing that struck me immediately, and I was appalled, was that they were given worn-out Corsairs. These planes had operated from the beach under difficult maintenance circumstances, had operated from the *Badoeng Strait* at one time, and the same airplanes had been in another carrier. They were old and not really very safe to fly. They were in great need of replacement. This disturbed me a bit and I immediately got in touch with commander, Fleet Air, Japan, and laid it on the line about these Corsairs that were so bad. Finally we did get some replacement Corsairs when people began to pay attention and get as concerned about it as I was, and the thing got better at that time.

We operated very successfully with United Nations Naval Forces, Yellow Sea. There were five different nationalities and it was amazing how well we were able to operate in very intricate maneuvers at night, reversing course and changing the screen at the same time, ships running around in the dark. The Australian skippers were real good and so were the British. They didn't mind getting out their whip and snapping it at the U.S. destroyers at all. But there was a tremendous amount of competition between the Australian and the British ships. They were each trying to outdo the other; they always tried to show the other one up, not apparent but in little subtle ways. It was an interesting thing to watch.

Finally, Capt. William A. Schoech was flown aboard in the Yellow Sea to relieve me. After the customary inspection of the ship, I departed, with orders to report to commander, Carrier Division 17, an antisubmarine warfare carrier division, as chief of staff to the commander.

15
Korean Knights

Captain Gerald G. O'Rourke

At long last, the day of departure for the Western Pacific finally arrived. We flew to Norfolk to join the carrier, the USS *Lake Champlain* (CVA-39), watched while our beloved Skyknights were hoisted aboard by a huge crane, and explained time and time again, just what the airplanes were, how fast they went, how heavy they were, and why they carried no bombs at all. The entire air group, four squadrons strong, flew aboard a few days later, and we started explaining all over again.

To the air group, we were true curios. They had completed a cycle of frantic training just as we had, save that most of their flying was done in daylight, and each of the two Banshee fighter squadrons had qualified fully in attack missions of bombing and strafing. This was to become their major role off Korea. They were excited and enamored with their bombing prowess, and picked at us mercilessly in the Ready Room, where we occupied a back corner and a small desk.

The Skyknights were brand new. There had never been a jet nightfighter detachment used in combat. All the previous deployments had used Corsairs, and had spent almost every mission in the role of night bombers. Everyone we talked to wanted to know what it was that we expected to contribute. We were going to intercept MiGs, we said. MiGs don't come any-

VC-4 Det. 44 flight crews posed by a Skynight at NAS Atlantic City before deploying to Korea in 1953. (Left to right, bottom row): *AT3 Ben Lataweic; AT3 David Lockwood; ETC Linton Smith; AT3 Pete Karnincik; AT3 Bill Hartsfield.* (Top row): *Lt. (jg) R. S. Bick; Lt. B. R. Allen; Lt. G. G. O'Rourke (OinC); Lt. G. L. Wegener; Lt. (jg) J. W. Brown. Bick and Smith were lost in a night engagement with a MiG-15, but were credited with a probable kill.* (Courtesy Capt. G. G. O'Rourke, USN [Ret.])

where near where you are going to fly, they answered, adding, "Why don't you carry some bombs and get into the real action?"

We became very disgruntled about the needling, and tended to become clannish. In private, we wondered ourselves just when we were going to find out about our missions. The Pacific Fleet had asked for the Whales. Obviously, someone out there wanted us for some definite reason. Effects had to have causes. Undoubtedly, there was a vast plan all ready to be set in motion as soon as we arrived. These matters always started with staffs. Some admiral wanted us. He had probably assigned his brightest, quickest, young commander to supervise our use. Soon now, this commander would get a message to us. Better yet, perhaps he would fly to meet us en route, and brief us and get us ready during the long voyage around half the globe.

In the meantime, we studied and trained, flew every few days, played a lot of bridge and poker, and read all the available "dope" on the Korean air war. Special Intelligence people had come aboard to brief us during the trip.

They scheduled a grand program, in which each squadron was exposed to all sorts of charts and maps, hints and tricks and advice on how to escape capture if and when we were ever shot down over North Korea.

The voyage from Norfolk to combat over Korea consisted of forty-seven days of tedium and anticipation, and except for six whole days of port visits along the way, almost devoid of liberty and flying as well. The actual introduction to combat for us hotshot jet nightfighters was a notably disappointing affair. As far as the stalemated Korean War went in 1953, navy carrier operations consisted of sitting in almost fixed locations at sea and flying massive daily sorties that kept the countryside subdued and docile. Night operations were unusually lucrative, but the problems were tough, and the attackers very limited in number. There was, in the navy area along Korea's eastern zone, little or no air opposition by day, and none at night. Accordingly, there was little for exotic, complex, costly, and large aircraft whose sole purpose in life was air-to-air combat in the darkness.

The air group was having a good time getting into the swing of daily combat operations. Each day on the line, the ordnance crews would start in the dark to load the planes with bombs and rockets and ammunition. Just before daylight, the first launch would roar off into the breaking dawn, heralding a long day of hard, almost continuous work for everyone aboard.

Because of the overwhelming air umbrellas that the navy kept over the eastern half of North Korea, and the air force and marines sustained over the western half, the living habits of the enemy had altered, just as any military tactician would have predicted. All their supply and logistics operations were conducted by night, or in weather that precluded air bombardment. Trains and trucks were hidden under camouflage by day, brought quickly to life as the sun set, and run hard under the friendly blanket of darkness throughout the night. The massive day strikes from carriers and shore bases found little to shoot at, while single night raiders would come upon juicy truck caravans and overladen supply trains almost every night.

We now realized what a hindrance it was, in this type of war, to have an airplane that was built for night fighting only and would not carry even a single bomb. The Corsair and Skyraider night pilots from the other carriers were having a ball, picking up a lot of medals, and doing a tremendous amount of damage. For these lucky fellows, train busting under parachute flares had become quite a sport. It wasn't for amateurs. The real pros used to go train hunting at every opportunity. A big incentive came from the medals and awards system then in effect. If you stopped a train and beat it up pretty well, you were a cinch for either an extra Air Medal or a Navy Commendation Medal. Neither of these, colloquially called "Boxtops" and

"Green Weenies" by much-decorated pilots, were any incentive, since each pilot had a chestful already. But if you could get a confirmed kill of a locomotive, then you were eligible for a Distinguished Flying Cross, which was a real medal. To get a confirmed kill of a locomotive, the engine had to be moved off the tracks. Just stopping it alone wasn't good enough. It had to be toppled over, or moved in some way to get the wheels clear of the tracks.

In the mountainous terrain of North Korea, each railroad line traversed a series of tunnels, any one of which provided surefire safety for an engine. Each engine, moreover, was very heavily armored to protect its boiler from machine-gun fire. These details set the stage for some of the weirdest duels in the history of armed conflict. A Corsair pilot, laden with flares, bombs, rockets, and 20-millimeter guns, would coax his overburdened plane into the night sky far at sea, leave the carrier astern, and struggle for altitude during his lonely journey toward the beach.

Straining his engine under high RPM, he would carefully navigate with his radar to arrive over the beach at an altitude that kept him clear of small arms and light AAA. Hunched over his radar and instruments, the pilot would avoid the known flak traps, avoiding radar-laying AAA guns who lay in wait for him. He'd finally arrive over a predetermined point, where intelligence had revealed the presence of a ChiCom supply route. Now he would start his stalking tactics, using instruments only for flight attitude reference. Cockpit lights were dimmed almost to darkness, allowing his eyes to accommodate to the inky soup outside. Warily timing his courses, he'd "travel the route" in search of telltale lights of trucks, or—best of all—the gray streak of locomotive smoke.

When he found trucks, he'd glide in, trying for the lead vehicle with a bomb to jam the column. Successful at this, the flare drop run came next. Then followed series of passes using a 20-millimeter to strafe the trucks, hoping for a secondary explosion to light the area better. This was fun, and was usually cause for calling in a buddy who might be running another route ten miles west. Working in relays, alternating flare dropping and strafing, real punishment could be meted out.

With a train, however, the tactics were quite different. The prospect of that DFC became paramount, and the engine was the only target—to hell with the boxcars! It was dangerous to get help, too, because your buddy might get the engine—not you. On the other hand, you had to have confirmation, so someone else had to help you out. The best bet was to try to hit not the engine, but the tracks forward of the engine, between it and the next tunnel. This required a bomb and a good drop, but with luck, there was enough time available. Next was the second rail cut, back between the rear of the train and the last tunnel. Another bomb, again carefully placed, could do this. Now the game turned into one of defender's ingenuity and pilot's skill. Machine-gun fire was useless for strafing ironclad engines. But they could be used against the railroad repair crews who would try to repair the track cuts. The rockets now became the vital weapons; a couple of good rocket hits on the engine, or under the engine, could topple it over, or twist it off the tracks. The flare supply was equally vital, since extreme rocket accuracy was required. A lucky pilot who was able to "move" the engine would try to save at least one flare, then holler for a buddy to come on over. Using the flare, the somewhat irate buddy would be shown the displaced engine and the DFC was ensured.

For us, the prospect of night combat was exhilarating, and formed the basis of our morale. We were the first jet nightfighters to operate in combat from a carrier deck, and we felt that there must be some opposition up there in the night skies. In desperation, we tried almost everything. We substituted for other planes on day carrier patrols. We struggled in orbits overhead in bright sunlight watching Banshees and Sabres climb high above us with ease. We ran escort missions for photo planes, but couldn't keep up with them, even when they slowed and we used max power. We ran a day road recce—about the silliest mission in the world for a huge, radar-loaded Whale, since our only armament was quadruple 20-millimeter guns.

We had other more mundane problems that got us into real trouble. The carrier deck was wooden. The Skyknight's tailpipes were canted downward to keep the exhaust clear of the aft fuselage. Even at idle engine speeds, the heat gushing onto the deck was formidable. At full power, such as when waiting for the catapult shot, the exhausts were like twin blowtorches, capable of turning steel plates into cherry-red griddles. There were no blast deflectors in those days, and the catapult crews had to scramble about on the protective steel deck plates aft of the catapults when each aircraft was readied for launch. After a Whale was fired, they couldn't get near the catapult tie-down area because of the cherry-red deck plate. In early shots, we bent and buckled almost every plate, causing a redecking almost daily.

Next a floating deck plate was tried, held in place under small lips, but we managed to break that one loose and sail it down the deck in the breeze. The next invention involved cooling the plate with water during the shot. We melted the cooling lines. Next came a fire hose, held by a man. The plate got hot, the water hit it, a cloud of salt steam arose and wafted aft through the pack of shiny Banshees, Panthers, and Skyraiders. Now the whole air group hated us for rusting up their paint and wax jobs.

When at idle for long periods of time—fifteen to twenty minutes while waiting for our turn to be launched—the heat would bake the wood, which had often been previously saturated with oil and fuel spills. Eventually, a minor conflagration would start. It wasn't serious, but it was still a "Fire on the flight deck" over the public address system, and the whole damn ship would rush to fire quarters. We could cure this by having a man handy with a CO_2 extinguisher, but as things go, this wasn't always done, and our unpopularity grew.

We tried parking the plane with exhausts extending over the deck edge. We burned up a few fire hoses and life rafts in the catwalks. We tried to spot the planes carefully—park them in places where little wood was exposed. We burned out the Babbitt in several arresting wire pulley sheaves. We changed the spotting plan. We burned out several landing lights. This was considered the best solution, however, since we were essentially the only night fliers aboard, and the loss of deck lights hurt only us.

The crowning insult, however, was our treatment of the ship's catapults. The F3D was a very heavy plane, and the ship's catapults, although hydraulic, were the last word of that breed. In later years, the steam catapults came into use and provided relatively simple cat operation at great power levels, but in those days, the H-8 hydraulic catapult was a Frankenstein of complexities, all designed to squeeze the last possible ounce of push from the system. The Skyknight, at its higher gross weights, often needed every one of those ounces and then some. Whenever we were fired, the cat crew would cross their fingers, since a miscue at that setting usually meant a runaway cat shot. This happened to us twice. The first time was en route to Korea, and the resultant damage to the forward end of the ship required repair by hotshot crews from the navy catapult facility in Philadelphia. They were flown halfway around the world to catch the ship and effected the repairs while we were under way.

The Whale, moreover, wasn't harmed a bit. Just as the catapult fired, the wire bridle dropped from its hooks, leaving the plane sitting in place, engines roaring, barely beginning to roll, while the shuttle of the catapult went screaming ahead at about a zillion miles an hour, dragging twenty-seven pounds of wire instead of twenty-seven thousand pounds of airplane behind. When it got to the end of its stroke, nothing could stop it. Wham!— into the stops, bending and warping all sorts of machinery and structure down below.

No one really ever found the trouble. Investigations were detailed and thorough, but inconclusive. The probable cause was blamed on the Skyknights, largely because they were a lot of trouble anyway. They burned the decks, took up all that room, needed special spare parts, ruined wax jobs, and didn't carry any bombs. Worst of all, none of us were maimed or killed in the incidents. Most fighters have enough thrust in such a situation to get moving pretty fast without the catapult. They are hard to stop, and go skidding off the bow with brakes locked. Not so the Whale! Its poor acceleration left it barely moving. A quick throttle chop, full brakes, and we were ready to turn around, go back, and try it again.

The constant battling for a decent mission and the rude shattering of our illusions about "our commander" made us all restive. Relations with the ship and air group were going downhill. We were an all-volunteer bunch, we wanted a piece of the action, and all we were getting were a few casual crumbs and a lot of "WestPac salutes" (arms extended sidewards, palms upraised, quizzically expressive, "Don't ask me"). Young, brash lieutenants, I soon discovered, were not to be heard, preferably not to be seen at morning planning conferences.

These daily meetings were scarcely more than rituals by this point in the war. Everything had already been done before a thousand times or more. Schedules, routines, targets, were all old hat to the planners. There were almost no more lucrative targets for daylight attack, and the byword was "High, fast, and once," in tactics discussions in the Ready Rooms. The greatest achievement of a commander, from squadron to task force, was to finish the tour and get home without losing a man. New ideas were as dangerous as new weapons, and I was full of both. Air defense of a carrier was just not considered essential. The Skyknight, as a new and better defensive weapon, was unwelcome. Off Korea, by 1953, a ChiCom air strike against the carrier force just wasn't credible. If they had wanted to do it, they'd have done so long before. Accordingly, the only real air emphasis was on attack, in support of the stalemated battle lines along the 38th parallel.

Ashore, however, there were marine air groups operating from fixed airfields with their carrier-capable aircraft. Although most marine squadrons were regular attack types, or fighter types being used primarily for the attack mission, one squadron of nightfighters, equipped with F3Ds, was doing a quiet, but magnificent job. This was VMF(N)-513, based at a USAF field called K-6 near P'yongtaek, about sixty miles south of Seoul. We had heard rumors, sea stories, and some details of their exploits, although their reputation was one of secrecy—or at least of little fanfare. They had five MiGs to their credit, we heard, each one bagged at night by a Whale.

U.S. Marine Corps aviation, nominally an integral part of naval aviation, but highly independent nonetheless, hadn't fared much better than the navy in the between-wars interlude. However, the marines had the foresight to

The first U.S. Navy jet aircraft designed solely as a nightfighter, the Douglas F3D Skynight was operated with great effectiveness by marine and navy flight crews in Korea. (Courtesy National Air and Space Museum)

maintain a small, dedicated core of World War II night-flying experts, from which several good squadrons were organized, assigned cast-off navy aircraft, and dispatched to Korea. VFM(N)-513 was one of these, notable for having transitioned from props to jets while in nightly combat. The squadron had exploited the F3D's unique radar capabilities in many ingenious ways escorting B-29 raids over North Korea. It had an excellent reputation and was held in the highest regard and esteem by its parent command, the First Marine Air Wing. They never scrubbed any mission for bad weather, pioneered all the successful night bomber escort tactics, and held the undying respect of the Russians, the Chinese, and the valiant USAF B-29 aircrews as the incomparable masters of the night skies over North Korea!

This was obviously where the action was—for us. We began to politic around with our bosses, asking for a short period ashore with the marines. We needed night flying. We needed instrument flying in weather. We weren't getting it aboard ship, and were rapidly becoming useless and cumbersome appendages to a fighting ship as a result. Our bosses agreed; then they disagreed. Then they agreed to disagree. This went on for several days as messages flashed back and forth from the ship to Japan, to the marines in Korea, and to their bosses at another field. Eventually we got there, even though the bosses were still disagreeing when we left the ship.

The second, and final, F3D catapult disaster came on the night when we actually departed the ship for our tour of duty ashore with the marines. It was a good night as far as weather went but was black as pitch, with no horizon at all. I lined up on the starboard cat, checked out, turned up, signaled okay, lights on, cat fired—and nothing but a thump under the nose wheel as the shuttle screamed under it. There followed a slow-moving sensation, as I started lumbering toward the bow. I waited a few seconds. Then I chopped throttles, jammed brakes, and was shortly taxied across the deck onto the other cat and fired off without further incident. About ten minutes later, while comfortably cruising towards Wonsan at 25,000 feet, I broke out in a cold, shaking sweat. The thought occurred that those few seconds of waiting, while trying to figure out what was happening on the catapult, had undoubtedly carried us very close to the darkened bow of the ship. We had missed a black, watery grave only by inches!

We had to stop after the road recce mission at a field on the eastern side of the peninsula called K-18, which nearly became our final destination. It was a scrubby place, fairly near the front, used by the navy as an emergency divert for wounded carrier planes and by the South Korean air force as a regular combat base. Every American there had a transient's attitude of unconcern. We landed with no further orders, just before dawn, low on fuel and out of ammo. The airstrip was covered with Marston matting, a net of steel, which chewed up our tires pretty badly. Refueling was an all-day affair, as each gallon was sucked from a huge pile of 55-gallon drums by a crazy portable pump. We spent several hours finding the communications bunch, flashed off some messages requesting instructions, and waited, and waited, and waited.

Orders finally arrived to proceed to K-6 to join the marines. Our departure from K-18 was full of thrilling fun. It was late in the afternoon, almost at sundown. The planes had been loaded with fuel. The wind and the wire matting runway weren't very compatible, and the length of the runway was greatly abbreviated by a sand dune rising about fifty feet in the air off its end. Anxious as we were to leave the transient world and get to the action, we overlooked that dune until suddenly there it was, dead ahead, as the plane was staggering along the matting desperately trying to gather enough speed for flight. We had to hold our breath, pop full flaps as we broke ground, balloon over the sand dune, and settle painfully close to the sea on the other side as we picked up speed.

The flight to K-6 took all of twenty-three minutes, in and out of clouds. It required about ten changes of radio channels as we were passed from one radar-controlled sector to the next. Our unfamiliarity with this luxury of radar following only complicated matters. On arrival at K-6, the sun was down and it was raining. Our heavy fuel load was still aboard. We had requested an hour or so of free flight around the area for familiarization. Instead, we shot approaches on radar in a rainstorm until we burned down enough fuel for a safe stop.

Taxiing onto the apron after landing, we were astounded to see a row of more than twenty black F3Ds, nestled closely together with folded wings. We were eagerly directed to open spots in the line and profusely greeted by a small crowd of pilots and ROs. Joining the marines was like coming home to mother at Christmas time. They were happy to see us. Our four shining airplanes augmented their resources. Our crews, even with their limited experience, looked good to them, and the future prospects of having some forty-five additional mechanics and technicians was manna from heaven for their hard-pressed ground force.

These were our kind of people! We loved them instantly! After an initial checkout mission with a marine radar operator, we navy types were fully integrated into the squadron. We were scheduled just like any other team. Our airplanes were intermingled with the marine planes, even though we tried to do most of the flying in them ourselves.

There was no doubt about a mission at K-6. They had their orders, were happy with them, and were marching off on their own to do the best they could at the task. This task, I soon learned, was twofold. The basic one, which was a continuous problem, was to keep two planes airborne over the waters just west of North Korea, under ground radar control from a lonely island called Ch'o-do, each night from sundown to sunup. An air force squadron of Lockheed F-94B nightfighters from another field near Seoul helped in this mission, on a fifty-fifty basis. Patrol duration was usually one hour on station—sometimes just thirty to forty-five minutes. The purpose of the patrol was to have nightfighters instantaneously ready to combat any ChiCom air attack.

In daylight hours, the Sabres from several fields took over the task, and had almost daily encounters with the MiG-15s over "Mig Alley" along the Yalu River. Antung, just across the Yalu, was the MiG base. Even though the Sabre pilots could see the field, the runways, and the traffic, they were not permitted to attack the field, since it involved a penetration of Chinese airspace across the river. At night, when we took over the duty, the same restrictions applied, even though it was far more difficult to navigate by radar with enough accuracy to remain clear.

For the night control, the ground radar station at Ch'o-do monitored and directed all aircraft. Crossing the Yalu, or penetrating directly northward from the sea, brought quick radio orders to "Reverse course. You are over the fence."

The second marine mission, in which the Skyknight had really made a name for itself, was to escort the nightly B-29 Superfortress raids over North Korea. This was a fascinating problem, involving an unusually high degree of coordination, crew proficiency, and radar performance. The huge B-29s came once each night from bases in Okinawa and Japan, flying long distances over water, or up through the Korean Peninsula. Their routes varied with each mission. Our task was simply to protect this stream of fifteen to thirty lumbering bombers from MiG attacks. Had the missions been daylight ones, most of the operation would be a visual affair. At night, however, it was all-radar, with navigation and timing of the utmost importance.

Like us, the B-29s flew only at night. In bright sunlight, their presence would have drawn a hornet's nest of MiGs and, despite the best efforts of the Sabre drivers, would probably have involved high rates of attrition. At night, they were relatively safe. Their high altitude kept them free of the light AAA and the ChiCom lack of any sizable force of nightfighters minimized that threat. Our task was to keep that threat minimized.

For VMF(N)-513, an escort mission involved eight to twelve planes at one time. One guarded each side of the bomber stream on the inbound leg. Two others loitered around the target area itself. Two more picked up the escort duty for the outbound run from the target. Two more might be required to patrol the outbound penetration leg, and another two to four drew the very choice assignments of barrier patrols, placed between the target and Antung to intercept any MiG defenders.

The operation was massive, to say the least. Radio channels were full of terse orders and reports. Because the entire operation took place over the enemy's rear support area, far behind their front line above Seoul, they had countless ground radars available for use. The raids presented very little surprise as a result. There were, first of all, very few lucrative targets to warrant the use of B-29s. The raids were nightly, one-shot affairs for the most part. The massive B-29s provided huge radar targets, which meant early and continuous tracking by enemy ground radars. Our nightfighters arose in mass numbers only once each night, which meant that this activity offered another good clue. There were plenty of North Korean infiltrators available for such spying. As each of us launched from K-6, a small signal light would flash from a nearby hill—on to the next, and so on. The enemy air defense net probably had the word within five minutes of takeoff time, while we were still climbing for altitude over South Korean territory. Any good ChiCom air controller, after watching the proceedings a few times in his underground information center, could estimate both the target and the time on target with uncanny accuracy—certainly in plenty of time to launch the MiGs, move the mobile searchlights, position the AAA guns, and alert the people at the actual target complex.

While the Ch'o-do patrol missions were considered milk runs, the escort missions were just the opposite. They required extensive briefings, a lot of study of charts and tables, extremely careful navigation, unbelievably intensive inflight coordination between aircraft, and a truly professional pilot/RO team. The whole squadron was reversed from that of a day outfit. Crews slept in daylight, briefed at 1600, flew a flight sometime between 1800 and 0700, drank booze and swapped hairy tales from 0700 to 0900 or so in the morning, then turned in to sleep until about 1500.

The maintenance gang was split into two shifts, each handling a twelve-hour segment. During daylight, from about 0700 on, one shift concentrated on the routine fixing of flight discrepancies, regular checks and inspections, and the heavy repair work, such as engine changes. In the evening, as the first flight was preparing to launch, the other shift would take over. Their basic job was to get the planes out, back in, refueled and rearmed, out and in again, and turned over to the duty day crews in the morning.

The Ch'o-do patrols, generally routine, could get very interesting from time to time. There was no set routine to the pattern for covering the area. In fact, we went to rather ridiculous extremes to avoid establishing any pattern. We flew at high altitudes, usually above 25,000 feet. There were generally two planes on patrol at all times, each flying independently. To avoid the embarrassment of a mid-air collision, one would stay below an agreed reference altitude, the other above. Whenever one wanted to penetrate into the other's allotted airspace, a quick radio call could clear the way.

We were quite sure that the ChiComs also had a rather large-scale radar coverage going. This meant that they could track, but not identify each of us fairly well. We were also quite certain that they had a well-integrated radio direction finder net tied into their radar control information center. Our own controllers, down in their lonely vigil on Ch'o-do would give us vectors, or ordered courses to fly. Instead of repeating these orders very carefully, in the most clear and concise manner, we would grunt a quick reply, then proceed to fly our own vector, which might be somewhere near, but never on, the course ordered from Ch'o-do. To do otherwise was folly. If MiGs were coming after us, any prior knowledge concerning headings, speeds, and altitudes was highly valuable to them, so why give it away free?

Our other danger was from several apparently mobile "one-shot Charlie" large-caliber AAA batteries. These would rarely waste a full barrage on us, since we flew far too high for them. They would, however, carefully track us from time to time, attempt to predict our track and altitude, and then lob off just one or two rounds, in hopes of a surprise hit. One-shot Charlies never really caused any trouble, but the occasional flash, close below, was rather startling on a quiet night.

The MiGs were our big concern—MiGs or bad weather. We'd never get both at once, since the MiGs just weren't very good at flying in clouds. Their planes were the essence of dayfighters. They were small, powerful, maneuverable, short-legged as far as range and endurance were concerned, heavily armed with good guns, and extremely fast by our standards. They were so much faster, in fact, that they could climb at a higher speed than we could dive! This strange disadvantage came from the sharp disparity in the wing designs of the two planes. The MiG had highly swept wings and could fly right up to the "sonic barrier" of Mach 1.0 before it ran into very heavy aerodynamic drag. Because of its swept wing, however, it needed a lot of speed to be efficient, a lot of speed for takeoff and climb, and a very high power-to-weight ratio to be maneuverable. The Whale, on the other hand, was built like a transport. Its straight, fat wings were great for takeoff, for turning, and for climbing at slow speeds, but they would really balk at any try to get above a speed of about 0.75 Mach. Even when going straight downhill, the intense wing drag would rarely let you get beyond 0.78 or 0.80, and at that speed the tail would quit working and you'd lose most control until you descended into much thicker air. In this thicker, warmer air, the true speed of sound would rise, the shock waves that caused the drag and the "wiped-out tail" would abate, and control would come back in with fierce suddenness. An unwary pilot could then get into serious trouble by pulling too many "gees" and overstressing the airplane. So, in terms of air-to-air fighting, we should have been hopelessly outclassed by the MiGs in daylight or a brightly moonlit night.

The Whale, however, had radar and the MiG did not. Our plane was beautifully stable in flight handling and the MiG was pretty touchy. We had a cockpit full of instruments and the MiG had few. We had lots of fuel and the MiG was always short. Last, but not least, the Skyknights had good, well-trained nightfighters in the cockpits, and the MiGs had a bare handful of pilots capable of this type of flying.

As a result, it was the Whale that roamed at will over North Korean airspace, and the MiG who was capable only of isolated hit-and-run tactics in futile defensive gestures. The night, the radar, the training, and the brains of the Americans provided the superiority. However, when one of our superior assets was missing, we were in deep trouble. Without radar, we were worse than useless—nothing but a fat target for their practice. A poorly trained or stupid pilot/RO team could get themselves bagged quite readily.

So, in daylight, we stayed home. This was our worst fear, since we were nearly helpless against any high-performance sweptwing fighter. Once he got us visually, he could out-climb, out-dive, and close in at will from our stern. Our only defense was to turn as sharply as possible, forcing him to overshoot. If he could keep us visually even while overshooting, he could pounce back in at will, and eventually get in a hit. So we stayed home in daylight.

Some of our fiercest pride in the Whale and in ourselves was that only three U.N. aircraft types were officially permitted to seek engagements with the MiGs. The USAF F-86 Sabre was, of course, the first. The USAF F-94B nightfighter was the second, and the marine and navy F3D was the third. All the others—the F9F Panthers and the F2H Banshees and the Air Force F-84 Thunderjets and the ROK P-51 Mustangs—had been assigned only to bombing and strafing and were restricted from seeking out the MiGs near Antung. The MiGs were more than a match for any of them. If attacked, of course, they had to defend themselves, but their geographical mission assignments, and the ceaseless barriers flown in daylight by the F-86s and at night by the F-94s and F3Ds, usually precluded inadvertent mixing.

A full moon is a boon to young lovers and MiG day pilots who had to go at night. To the night types in Whales the full moon was a treacherous siren, lurking you into flying about the cloud tops, where the big black or blue hulk of the Whale became a beacon in the sky, a moving dark blob before a rolling, static white background.

We avoided the moon like the plague. If there were weather piling up to 27,000 feet, with a moon of any sort atop, we'd fly at 26,000 feet, buried in the crud, hidden from the searching eyes of any MiG pilot, yet just as all-seeing with our own radars. When the clouds weren't stratified that far up, but lay about in billowy piles of cumulonimbus clouds, we'd dart back and forth among them, clipping in and out of corners, changing course often to avoid both a steady track and the more severe storms.

We had two sure-fire safety relief valves. One was bad weather over Antung. This happened quite often, and we never saw a MiG on any such occasion. They flew only when their weather was extremely good at their home base. The reason was quite understandable. They flew using day tactics, in tight formations, with little fuel reserves, and probably with few navigational aids to help them find home. On top of all that, they had relatively poor instrumentation, and a nervous airplane to begin with. Moreover, their pilots just weren't long on such experience and probably just wouldn't go in bad weather.

The second respite from MiGs came during those hours of the night after the MiG team had tried to zap one of the F3Ds. We didn't figure this one out for several months, but finally came to the conclusion that they had only one team of four MiG pilots who would (or could) fly at night in combat. Whatever their union rules were, they only tried the gambit once each good night. If one F3D had been played with early in the evening, the rest of the night was sure to be no challenge at all.

The gambit they used against a lone Skyknight was pretty crude, but effective. They had no airborne radars, and had to rely completely upon their ground radar controllers to get them into visual range behind us. Because the Whale changed course so often, and wasn't very visible to begin with, and mostly because it had a set of radar in its tail, they had to come in fast, and come in very accurately. They tried this many times with absolutely no success. Either the Whale crew would spot them climbing out with the nose radar, and start after them, or the tail warning radar would signal an alert, and violent quick turns would spoil the whole show for the MiGs.

One of their first mistakes was to fly at a nice safe, slow speed—one that made it easy for the wingman to stay in position on the leader. After a couple of them had been zapped by Skyknights who "snuck up" on them, they got wise and pushed their cruise speed up above the maximum speed of the Whale. This made them almost invulnerable to an attempted stern attack, but left them open still to a modified type of curving lead pursuit attack from one side. This maneuver was damn hard for even a good F3D crew under ideal conditions. A poor crew never had a chance, and would wind up far astern, watching the juicy targets draw away from him.

Another trick the MiGs tried was to lure a Skyknight into a trap. This was the old red herring trick. They would set one MiG up ahead of a Whale, at a comfortable speed, and slow him down until the F3D crew spied him on radar. The F3D crew would get excited and try to catch the lure. In so doing, they would tend to forget about the tail. At this point, while the pawn up ahead was cruising along just out of gunfire range, three MiGs in formation would come whistling up the Whale's stern, in hopes of catching him visually and getting off a few lethal 37-millimeter rounds. If the Whale's crew smelled a rat early, and carefully watched the tail radar while chasing the bird up ahead, the oncoming MiGs astern would show up like beacons, and a tight turn would throw the party into the wildest type of confusion. In all the games we played those nights, only one MiG, one F3D, and possibly a second F3D were actually shot down.

Another variation of the red herring was to draw a Whale through the searchlights that made a bright tepee over the B-29s' target each fair-weather night. The lights were radar controlled, and would all zero in on one point in space. The MiG pawn would try to get the Skyknight to pick up the chase, then fly through the apex of the lights, where another MiG was supposed to pick him up and shoot him down. The ChiComs, or Russians, whoever it was who was directing the show, were still living in World War I, when searchlights were big time against 70-knot airplanes at 300 feet. Against jets at 20,000 feet, the problem just wasn't the same, and the searchlights were totally useless—so useless, in fact, that one sharp marine crew had turned the tables of the trap by dropping the chase after the first pass through, then loitering for a minute, then doubling back around to pick up the MiG as he started through another time. Scratch one MiG, and that was the end of that. The MiGs didn't play that game again, but the ground types never gave up. They also tried relentlessly to snare a B-29 in the lights for assistance in their AAA problem. As far as I know, they never hurt a single B-29 either.

Navigating and getting in and out of K-6 during a heavy rain was a real challenge. On takeoff, once clear of the hazards of the slick runway, adjacent mud flats, and a nasty hill off the end of the runway, the navigation problems began. Low-frequency homing radio stations were the major aids, and they weren't too reliable when the thunder and lightning started. We'd use them, for whatever they happened to be worth that particular night, and try to use the radar, and try to navigate by time and distance and speed, with an occasional assist from air traffic controllers in the area.

Coming home was another affair altogether. Once we were through dealing with the Seoul traffic controllers, we'd track inbound on low-frequency homers until GCA could find us and start giving us some real assistance in getting onto the runway. The GCA outfit was a great bunch that did their level best with what they had—which wasn't much. Instead of the best gear in Korea, we had the most ancient decrepit GCA equipment on the peninsula. The reason was fairly simple. K-6, officially, was managed by the air force, yet housed only marines. Whenever a good piece of gear arrived, it was sent first to an air force field that housed air force squadrons, and their castoffs were sent down to us. Yet the air force boys out there in the GCA trailer, perched atop a mud hill next to the runway, were the finest in the world. If an approaching plane fouled up a landing and skidded badly, they, personally, in their trailer full of electrons, were a likely target. They never ran, even when they knew trouble was coming. They tried. They tried hard.

But they couldn't beat the rain with the old gear. It just wasn't sharp enough or powerful enough to show them the difference between an airplane and a mess of raindrops. As a result, when we needed them the worst, they were useless to us. We had to help ourselves. And that we did. It became apparent, after a few GCAs, that our own radar in the airplane could look through rain with a lot better results than theirs. We could see the outline of the land, around which the river curled, and estimate where the runway was from the river bend. It was a risky proposition, since Korea was full of river bends, but if we started early, way out past the radio homing stations, and planned the approach as though we had no help at all from GCA, we could do a fairly good job of getting lined up with the runway. Once lined up, and with ranges available from the radar, we could ease down our altitude in steps, and eventually pick up the runway lights dead ahead. Once they were in sight, the problem changed from an instrument one to a visual one.

Now the name of the game was to get the heavy airplane down out of the sky at a slow enough speed to set it on the runway, in a good lineup, so that the marginal braking down the length of the runway would give us a fair chance of getting stopped before we became a deep-mudder off the far end.

After many weeks at K-6, with little or no word from the ship about future plans for the detachment, I took it upon myself to bum a ride over to Japan and to catch the ship while it was in port in Yokosuka. After making my way down to Yokosuka, and eventually to the ship, I was greatly disturbed to find the communicator looking for me with a very questionable message. It was from the marines at K-6, and asked simply for the service records of Lt. (jg) Bob Bick and Chief Linton Smith. While it could have been a routine administrative query from the detachment's yeoman, it could also mean a missing airplane and a lost crew—a lost navy crew—one of my crews. I immediately went in search of the intelligence people, trying to find out what had happened back in Korea since I'd left some twenty-four hours earlier. It was then that I realized how vast and impersonal a war can be. The navy knew nothing of either operations or possible losses of shore-based aircraft. That was the air force's domain. No one on the ship knew a thing. Finally, by appealing to a classmate from the Naval Academy who was then holding down an intelligence billet on the admiral's staff in Yokosuka, I discovered that there just might have been a missing F3D from the previous night's operations over the north.

I never did see the people I wanted to, but instead had my orders quickly stamped, and set out for K-6 once again. When I arrived, after another sixteen hours of agony in C-46s and DC-3s, I learned the worst: Bick and Smith were in fact lost. They had not been heard of since halfway through a routine Ch'o-do patrol on the night I was fighting my way across Japan in a miserable C-46.

It took about a week to reconstitute the event as it probably occurred. The week had seen an extensive search flown over the water areas near Ch'o-do, under particularly poor weather conditions, without any trace found of either the aircrew or the airplane. They had been under Ch'o-do control, at about 0200, flying what was supposed to be a very routine patrol. No MiGs had been sighted or reported by radar, when Bick suddenly asked for information on a radar contact. The controller at Ch'o-do had a difficult time even tracking them and held no contact close by. Bick had asked for permission to fire. He had been given a "Wait" for an answer, while the distracted young officer on the Ch'o-do scope pleaded for some guidance from his senior. Bick asked for permission to fire time and time again, but could get no clearance from the ground. Then he reported that he had fired, but had also "taken a couple of thirty-sevens," and was heading for home. He was THEN given clearance to fire. He disappeared from the scope while headed toward the land in the south, and all radio contact was lost simultaneously.

While this would seem quite a mystery to many, we in VMF(N)-513 were fairly sure what had happened. They had been suckered into the red herring gambit. A MiG had appeared in his scope up ahead. He had chased it furiously, trying to close to firing range. He had even reported seeing the tailpipe of a MiG, although we tended to ascribe that to his fertile imagination. He must have seen the other MiGs closing in on him from the rear. Chief Smith, the RO, was as conservative and cautious as Bob Bick, the pilot, was freewheeling and incautious. We had counted on Smith to exercise restraint for the team. He certainly would not have overlooked his tail-warning radar in this kind of chase. He had probably tried to get Bick to break it off and come around for another kind of a try in the melee. Bick, however, was probably of a vastly different mind. Here was his MiG, at long last, right up there ahead of him, almost in his gunsights. Break it off? Never!

And so he had held on, and so he had been hit. The plane may have burned. They may have been hit anew, or the cockpit may have flooded with smoke and electrical fire. We were never to know, just as we were never to know even where the plane had actually struck the water. The Ch'o-do controller just could not tell us.

Losing one of our five crews was a real blow. Until that time, we had developed a sense of immunity to tragedy. Certainly the loss of Terry and O'Neil back at Atlantic City had hurt, but that had been too early in the game to make a lasting impression. Bick and Smith had come all the way with us. The chief had been "Big Daddy" to all the young radar operators. They had turned to him for advice, both professional and private. The youngsters began to say to themselves, "If the Chief got himself bagged, what hope is there for me?"

One of them began to crumble under the strain. He stopped laughing and joking with the others. Humor just disappeared from his life. He smoked incessantly, and started to stammer when he spoke. He got on the nerves of his pilot, who complained mightily that things were going from bad to worse. Each succeeding hop for this pair became a mutual ordeal. They had lost faith in each other, and no more fatal sin can happen to a nightfighter team. The young RO finally quit. He stammered out a lot of words in a highly embarrassed state, but the fact was that he just quit.

I had seen a few aviators quit before this time, but never had been closely associated with the painful details. Here, however, I was the final resource for him—a lieutenant who was ten years his senior, and to whom such matters were largely alien. In later years, there were to be many similar situations, but this was the first, and therefore the most memorable for me. Then, too, he wasn't a trained pilot. He was simply a very young, very scared, technician. He had not been to college; he wasn't commissioned. His total pay was something like $150 a month. Almost all his peers did not fly, and did not expect to be called upon to fly. In comparison with the young pilot, he had far less "face" to lose by his action.

Yet his action undoubtedly was to mark him for life. Any man can be scared. Most men can struggle through a period of fear and survive unscathed but wiser. When a man is scared and must publicly admit the fact, some portion of his innate manhood is forever destroyed. Five, ten, or twenty years later, he will be faced down in some other situation for the want of it.

The loss of our crew stirred up memories of us for those back on board ship. Until this time, our navy shipmates had seemingly cast us adrift, assuming we would never return from our marine haven ashore. The full investigation of the combat loss had indicated that Bick and Smith actually had fired at a MiG, and had reported hits on the MiG. On the basis of this, they were posthumously credited with a MiG kill. This was the morsel of news that made us all famous back aboard ship. There, three jet fighter squadrons had been used almost exclusively as bombers. They never were allowed to make the long haul over to MiG alley near Yalu, and only rarely ever even saw an unfriendly airplane of any type. Almost every mission was, for them, a straight-in, "high, fast, and once" run on some target, into which they unloaded their bombs. On other days, some of them ran road recces on which they would strafe up the various cow paths and mountain trails that ribboned the east coast. None of their targets were lucrative in any sense—an occasional truck or ox cart, more often just another beat-up shack at a road junction.

Our supposed "MiG kill" captured their fancy, and I was called back aboard to report on our experiences. I flew down to K-3 in a Whale, then climbed aboard an ancient torpedo plane that had been converted into a five-passenger, carrier-capable transport. After a thumping landing aboard, I was greeted with great interest as a man from another world. I had worn my best marine fatigues and highly polished flight boots for the occasion, but they looked pretty sordid and scroungy amid the sparkling cleanliness of the ship. At chow I noticed the tablecloth, the large napkins, and the iced tea and salad for the first time. The same ship that had seemed so dirty and old when I left, now gleamed like a surgical operating room.

Explaining our operations ashore was a tough problem. It was so different from the carrier life that no one really seemed to understand. They tended to visualize it as far more trying than it actually was. In a short time, I realized that while we were heroes of a sort and very much a part of the ships's air group in spirit, there just wasn't any room for us back on board, nor would there be any better mission than the useless patrols over the carrier should we ever actually get back aboard. With greatly mixed emotions, I learned that the catapult had again failed to properly fire—this time with a propeller-driven Skyraider. The problem, then, was not with the F3D alone. Unfortunately, it had cost the life of a crewman to discover this, since the Skyraider had plunged into the sea.

After the luxury of a hot shower, and a blissful sleep on a large bunk between clean sheets, I left the ship early the next morning, retraced my route through K-3, and was home at K-6 in time for the 1600 briefing. The world of K-6 looked far more comfortable to me now. Here I lived, here I understood and was understood, here were my men and my faithful airplanes. Here was my home, and happy was I to be back.

It took the ChiCom MiG crew about three days to digest their kill. Once they had figured it all out, they came back to the fray like bloodthirsty tigers on the scent of wounded game. From then on, their forays came nightly. Even when the weather at Antung was less than perfect, they would sortie out in search of game. In VMF(N)-513, we began to feel the increasing pressure, and tried to devise ways and means of combatting them. Ch'o-do just wasn't very helpful. We held their controllers at fault for the F3D loss, which was probably very unfair. The controller in that case could not have known much, but he was not working with the world's best radar, either, and the melee had occurred a goodly distance from him. But the questions persisted. How could the MiGs foray that far south from Antung without making some kind of a mark on the Ch'o-do radars? We never got any advance warning on them from the ground.

The F3D's tail-warning radar suddenly became all-important. Prior to this time, when on a Ch'o-do patrol, if the tail-warning radar quit after getting airborne, most pilots would chance the loss, and proceed with the mission, using their forward radar, but trusting in the Lord for protection from their rear. After the loss of Bick and Smith and the earlier loss of a marine crew under completely unknown conditions, we decided never to cross the bombline inbound without a good tail-warning set.

A curious pattern quickly became evident. Once—and only once—each night, a single F3D crew would return to debrief an incredibly wild tale of combat. The tactics were much the same in every case—a target ahead, a short chase, and three targets suddenly closing from the rear. Tight turns, a melee by moonlight, targets and contacts flashing by both ahead and astern, frantic calls to the other F3D to come over and help out, and an eventual withdrawal by both sides. Some crews claimed they saw the red balls whistling by. I don't know. We never had a plane return with combat damage.

We continued to escort the B-29s, and ran across the MiG team there as well as to the south over Ch'o-do. Yet, never in both places on the same night. The evidence pointed more and more to just one well-trained team of four or five pilots. Were they North Korean? Hardly possible, since they could rarely fly even in daylight. ChiCom? Possibly, but again, their experience level in daylight seemed to rule this out. Russians? Most likely. There had been fairly positive evidence of their participation in daylight as an instructor cadre. The war was nearing three years of age. Many Russian MiGs were in use. New pilots had to be instructed. And the final bit of evidence, which convinced us of their nationality, was that, "If I were a Russian MiG instructor pilot, I'd be doing exactly as they were!"

The night they selected *me* was a good, coal-black one. I had drawn the select job of northern flank for the B-29 raid, from the turning point to the target. The plane I had was a brand-new one, which had been sent to us from a small stock of spares in Japan. Unlike the older planes, it had only two oxygen bottles instead of the normal three, a change that was to prove disconcerting for me, since I was not aware of it, or had forgotten about it at take-off. We had routinely used 100 percent oxygen on all night flights, for several good reasons—better eyesight, less chance of regulator foul-up, and a better chance to breathe if we were hit or hurt.

Ben Lataweic, my RO, and I arrived over the inbound turn point right on schedule, about two minutes prior to the projected arrival of the lead B-29. We were well above their flight altitude, and had just started a turn toward the reciprocal heading to pick them up, when Ben started getting excited. He had a contact. It was not headed toward us, but away from us, so it couldn't be either a B-29 or another F3D. The B-29s were well below and headed toward us, and no other F3Ds were supposed to be in this area. Just as Bick had, I began to smell my MiG, and poured on the power, turning and twisting on commands from Ben. I had queried Ch'o-do about friendlies in this area, but felt that any answer would be meaningless, since we were well into the north, deeply overland.

As I made turns this way and that, I suddenly became aware, out of the corner of my eye, of a light flashing briefly and erratically out ahead of us. At first, I paid no attention. The Whale had a lot of glass and plastic surfaces that were notorious for casting strange reflections back at the pilot. In a flash, however, I suddenly realized just what a patsy I was. The plane out ahead was a MiG, and he was so damnably confident that he was flashing his taillight off and on, just to keep me chasing him! As I realized this, I leaned over toward Ben's side, threw a glance at the tail-warning radar, and almost died of shock. There on the scope were six separate and distinct blips. Each pair signified one airplane, so that meant three of them closing in astern!

I muttered, "I don't like this, let's get out of here," and Ben came up on the intercom with the same conclusion. He had been so engrossed in the pursuit of the MiG ahead that he had momentarily neglected to watch his tail on the small scope. When I leaned over, he evidently glanced up at it. His voice rose a good three octaves as he shouted, "Hard port!"

Too late. I had already turned hard starboard, pulling as many "gees" as the plane would take. Around we went. The tail contacts stopped closing and moved off the scope to starboard. The forward contact also disappeared. We turned about 180 degrees. Suddenly, a huge blip flashed across the forward radar at less than a mile, quickly swinging clear to starboard again. I chased it around. Two more contacts on the tail. A hard turn to port. Another ahead, but lost in the turn. Another astern, a single now. Altitude dropping somewhat, but still above the B-29s. Silly, they've already passed well clear. We've been at this for a good five minutes, and we started to the north of them. Another nose contact. Perhaps a chance at him. He opened easily, outracing us to the south. Another on the tail.

And so it went, for a solid twenty minutes. We went around and around, milling about like blindfolded wrestlers, catching a hint of a target here, then there, then having one spy us for an instant. Altitude kept lowering as the high "gee" turns took their toll. We were a little worried about fuel, since we were far, far from home. We thought about hollering for help, but realized that another F3D would be no help; it would only clutter up our numerous contacts anyway, and make our firing problem worse. We might be shooting at a buddy.

We tried to just plain quit and head for home, down to the southwest. We steadied up on course 220 degrees and a new contact appeared astern. This time we played it a bit cooler, waiting to see if he closed. If he didn't, he was no threat. He did. He was a threat. We turned again, harder than ever—a complete 360-degree turn, trying to catch him afterwards. No luck—and no oxygen now, either.

Damn this new airplane. It hadn't been hot-rodded, marine-style, so it was no great performer. I wished we had an old one, which maybe would fly a few knots faster. Level out on 230 degrees, no contacts, head for home, drop down to a safe oxygen altitude, below 20,000, preferably lower, get the navigation gear cranked up. Don't really know where the hell we are, except that home is to the southwest.

Fuel low, but okay, we hope. No more contacts. They have far less gas than we, I remember, so we probably chased them off for home. We weren't much as far as heroes go. No MiGs to show. Scared, yes, but panicked, no! Sure would have been nice to stick a few rounds into one of those bastards. There's the navigation homer. My God, we can't be that far north! Yes, we can, it developed.

Now we knew what happened to Bick and Smith. Now I knew why the marine plane just disappeared. Poor tail radar, perhaps, and quick closure and kill by the three MiGs. Well, there are compensations. Now I had been shot at in earnest. At least I suppose I had. I never saw any red balls floating by, but then, I wasn't looking hard at the time. They were surely close enough to try a few rounds.

Was the mission a success? By what standards? Did we make the night safe for the B-29s? Yes. Then a success it was. Did we get a MiG? No. Failure. Did we chase them away? Well, you could possibly say so, but they didn't exactly run away and hide when we turned on them. Had we made them less cocky? Probably yes. The tail warning made the big difference. Without it, we'd have been bagged for sure.

We had daily rumors and confusing news briefs about the war's end. One day it would seem imminent—the next, remote. All the world, save us, seemed held in temporary paralysis while the crazy diplomacy was going on a few miles north at Panmunjom. For us, the war went on night after rainy night. As the news and rumors of a cease-fire mounted, the crews started counting missions, scrambling to be scheduled in order to qualify for an air medal or commendation ribbon.

Once the real war ended, the friendly dogfights began in earnest all over the Seoul area, where most of the airfields lay. They were mock, of course, but anything went. One field had a crazy bunch of South Africans, who flew "loaded" F-86 Sabres, and who knew that they soon had to turn the planes back to Uncle Sam and go back to routine flying in some old cast-off Brit types at Cape Town. They acted as though each flight was to be the last for them, and would jump anything that moved in the air, be it fighter, attack, or transport, and buzz it continuously until the pilot pleaded for mercy.

Everyone, it seemed, wanted a ride in the F3D. The jets were still in rather short supply, and were really curiosities throughout most of the world. A jet fighter normally carries only one person, which ruled out most freeloaders. The F3D, however, was the ideal answer for the VIP or for the mechanic who had always wanted to ride in a jet fighter. We would strap them in the starboard seat, make sure they had a barf bag handy, and blast around the heavens for them until they felt ill and wanted only to get the hell out of the smelly bird.

With the end of the fighting, and prior to the onset of the bureaucratization of everything in Korea, there was a lot of fun to be had. We flew "every airplane, every pilot, every day," as a general rule. This system kept everyone in good shape. We thought we might get back aboard, so we started concentrating on bounce hops. Most of the mechs in the detachment had labored over, but never flown in, the F3D, so we took them for brief rides. We performed regular training of all sorts, and even worked up a very modest stunt team. The marine F3Ds had been painted black, with red numbers. Ours were blue, with white numbers and letters. When we got our four-plane aerobatic show going, we dubbed ourselves the "Black and Blue Angels," hoping that the true Blue Angels would understand, back there in Pensacola.

I learned a lot of true respect for the Marine Corps during that tour. There is very definitely something about a marine that just doesn't show up in the other services. Some call it mule-headedness, others say dedication, still others say it's just lunacy. Whatever it is, I'm quite content with marines as they are, because this country just couldn't exist for very long without them. It makes little difference whether they be highly skilled technicians, center-city high school dropouts, or naive farm boys, their spirit of duty is ingrained to a point almost beyond the comprehension of other professional military men.

In aviation, the marines have usually been forced to grasp for straws. They were rarely allowed to buy new airplanes before the Korean War.

Instead, they flew whatever the navy had either cast aside as obsolete or unworkable. The marines made the Corsair famous after the navy had almost despaired of its use in World War II. Later, when the navy twin-engined F7F fighter proved unworkable on carrier decks, it was given to the marines and used well by them for many years. When the navy's grand experiment with the Skyknight resulted in a slow, fat, unlovely, and unloved monster, it, too, was passed along to the marines—who made it into the world's finest nightfighting machine at the time when nightfighters were sorely needed.

=16=
At the Brink
The Cuban Missile Crisis

Admiral George W. Anderson

When I came back to Washington from command of the Sixth Fleet to become chief of naval operations in August 1961, it was quite apparent that the United States as a whole and the navy in particular was very disappointed with the outcome of the Bay of Pigs. It had been a matter of continuing concern, and I personally sensed that there was a very high manifestation of the competitiveness of President Kennedy and Bobby Kennedy to make up for what had gone on. As a matter or fact, as soon as I became a member of the Joint Chiefs of Staff, it was clear that there was a weekly meeting at the higher State Department policy level as to what could be done to redress the situation. Indeed, solicitations were being made quietly as to any idea that could be offered to provoke Cuba into giving the United States an excuse to take appropriate action. These meetings continued, and then in September 1962 there came an increasing number of reports voiced particularly by Senator Kenneth Keating of New York about the buildup of Russian forces in Cuba, including missiles.

We in the military watched these and read his reports and were more or less sympathetic toward what Keating was saying. On the other hand, the actions of the administration emphasized—and remarks by administration spokesmen emphasized again and again—that they were only defensive weapons that had gone into Cuba. At the time, we were trying to get surveillance of Cuba through our U-2 activities and of course from other intelligence sources. There was somewhat of a hiatus in U-2 flights and, therefore, it was not until sometime in October that it was disclosed that indeed the Russians had put in Cuba surface-to-surface missiles. Well, you might say that Washington went immediately to general quarters and for a period of days the Joint Chiefs of Staff were meeting regularly. Gen. Maxwell Taylor, the chairman of the Joint Chiefs, was meeting with the special group and would come back and report to us what the special group wanted in the way of information and what their opinions were.

But the other members of the Chiefs felt that for one reason or another, partially perhaps because of General Taylor's hearing impediment, we were not getting the full story, as it were. Perhaps he was so busy with a great variety of things that he didn't realize that he wasn't telling us everything, though he thought he probably was. In any event, we in the Chiefs deliberated at great length on what the options were, what could and should be done, and there was, admittedly, a variation in the actions that should be taken.

At the same time, there was a tremendous buildup of military forces in the environs of Cuba and south Florida and in the Caribbean. We in the navy were reasonably fortunate that we had forces available at that time to make redeployments and handle them well and be prepared for any one of the contingency situations that had been voiced or considered in any way by the Joint Chiefs of Staff, or were being considered by the special group. We were able, quietly at first and then more dramatically, to get ships to sea and move airplane squadrons and, particularly, marine battalions. We had alerted marines from the West Coast, from the Pacific Fleet, to be able to move to the Atlantic coast. We had reasonably accurate timetables of what each course of action or each plan would require in the way of time. Preparatory steps were being taken because we didn't know just what course of action would be developed.

There was, of course, the idea of a quick air strike, advocated more by Gen. Curtis LeMay. There was the idea of perhaps an all-out invasion of Cuba, a question of blockade—we didn't use the term "quarantine" in the initial stage. We recognized that the decisions that were to be made were going to be made by the president and we hoped that the input of the Joint Chiefs of Staff would be completely and adequately considered. That, of course, had to be presented by General Taylor, although later on in the process we did have a meeting with the president and the entire Joint Chiefs of Staff, during the crisis.

We had very good intelligence reports at every Joint Chiefs of Staff meeting, particularly during the September and October period. The Defense Intelligence Agency regularly presented photographic intelligence that was available to us, and, of course, we had the special intelligence that came in. But, in any event, this whole development was being followed with intense interest from September right on through October. There were the usual Monday meetings with the secretary of defense and the Joint Chiefs of Staff, much of which was devoted to the Cuban situation.

But I had the feeling that, especially because of the competitive attitude of the Kennedy brothers after the disaster of the Bay of Pigs, that they would press this particular crisis to a conclusion successful from the standpoint of the United States. I sensed that competitive attitude on their part from what I'd read of their whole life, the actions of the Kennedy family, and particularly that they didn't like to lose—and they had lost in the Bay of Pigs. And we knew they were soliciting excuses for the United States to do something all during the course of these weekly meetings of the special group that would consider what should be done.

I personally felt that we should use this particular crisis to solve the Cuban problem. That, in my mind, was a primary consideration. We'd had Castro come in, we had a Communist-Marxist government established ninety miles off the coast of Florida, we had increased involvement in Cuba in the Soviet Communist empire and we saw it—I saw it at least—as a portent of further involvement of Communist domination in Latin and Central America, the threat to the Panama Canal, control of the Caribbean, and that threat should be removed as rapidly as we could.

A second consideration of mine of particular concern was the security of our people on the base at Guantanamo, the security of that base, the importance of retaining that base for the navy for future years. I would have liked to have seen, first, a dramatic buildup of our forces in Guantanamo Bay coupled with the removal of civilian dependents, unnecessary civilians down there, and second, a massive invasion of Cuba coupled with a massive propaganda campaign of all sorts—leaflets, radio broadcasts—to get the support of the Cuban people over the Russian Communist domination of the Cuban Castro government. And I would have followed up the invasion, which would have been amphibious and airborne, with a large civil defense program, medical aid, food, and so forth, and reestablish forthwith a free Cuban civilian government. I don't think I espoused that actively enough and I also did not feel that this thought was adequately put forth by either the chairman of the Joint Chiefs of Staff or Secretary of Defense Robert McNamara. However, we were prepared to do whatever we were told.

There was a reflection of another attitude, as pronounced by Adlai Stevenson at the United Nations, when he said on one occasion why didn't we just give them back Guantanamo, it wasn't of any value to us anyway. I think that was his peculiar point of view. I never saw any indication from higher authority, the president and the secretary of defense, that we would give back Guantanamo forthwith. That would have been sort of succumbing to blackmail when it reached that stage, and there was not that attitude.

Finally, the Joint Chiefs of Staff met with the president, and he said that he understood that there were differences of views of the Joint Chiefs of Staff about what he had decided to do, but he had decided to take this initial step of quarantine, but this did not preclude other actions that might be necessary if that didn't work. As we left the White House after that particular meeting, the president said to me, "Admiral, this is up to the navy."

I said, "Mr. President, the navy will not let you down."

So we had gotten in the position of moving our forces around to the best military posture the navy could adopt, and we had a pretty formidable position at that time. For example, we had all our available Polaris submarines at sea, we had a line of defensive submarine barriers across the Atlantic, we had our patrol squadrons scouring the Atlantic from the best available bases for reconnaissance of ships, particularly for Russian submarines. We had our marine battalions in a high degree of readiness. We had marine battalions embarked at sea on the West Coast, and some of them by that time were moving from the Pacific down toward the Panama Canal. In all, we had some eleven marine battalions committed in varying degrees of proximity to Cuba, varying from on the scene in Guantanamo to those moving down from the San Diego area down toward the Panama Canal.

When we were establishing the so-called quarantine line, from a naval point of view, I wanted the line far enough out so that land-based planes, especially the fighter-type planes that the Cubans had, could not interfere with the surface ships. In other words, we would have a measure of security wherever the line was established. On the civilian side, the idea was to move the line back toward Cuba, to give the Soviet ships that had been steaming toward Cuba more opportunity to turn back—in other words, to delay a confrontation. And so, actually, I think the line was set up at about five hundred miles, which was not too bad.

I wasn't particularly concerned that they were going to really damage our ships, but it would have made our whole operation a lot more satisfactory, manageable. We didn't have to put fighter cover over all these destroyers, for example, out on the line. It would have been better if we'd had it where the navy had proposed, but it wasn't a matter of vital concern, as far as I was concerned. They wanted to delay any confrontation so that the word could get back to the Soviet Union that they'd better call it off and then, in turn, get the word back to the ships.

I was concerned about protecting our civilian dependents because that is a prime responsibility of our service commanders all around the world. On the other hand, I was fully aware of the fact that you mustn't do this prematurely if it's going to affect the actions of the other side. We didn't want to give a sign of weakness or an indication that we were going to give up Guantanamo prematurely, and yet we didn't want dependents in there at the time any guns might be fired to start with. We worked that out very well, and actually this whole part of the evacuation at Guantanamo did not receive too much publicity because we did it swiftly and it was well organized; we used the ships and the airplanes that had brought in additional reinforcing marines to take them out. Afterwards, I also insisted and was successful in getting the dependents put back as soon as it was possible to do so.

I felt that we had an opportunity then to get rid of the intrusion of communism on a massive scale that had taken place in Cuba, and, therefore, prevent it from spreading in other areas of Latin America, and, although we didn't realize it at the time, the extension of Cuban intervention later in Angola and Ethiopia. This is what I tried to emphasize to McNamara personally, but he was really preoccupied with something else, I guess, at the time and I didn't get it across to him. What I would have done is pose a major challenge to Cuba and a secondary challenge to the Soviet Union. In other words, if the Soviet Union kept out, we were not going to attack them, we would let them get out afterwards, but we were going after Cuba, as such, and I don't think under those circumstances the Russians would have interfered if we'd done it in the right way.

Admittedly, from the Joint Chiefs' point of view, some of the sensitive negotiations, exchanges of information between the White House and the Soviet Union, were not filtering down to the Chiefs. That was so tightly held—maybe they gave it to Taylor and he didn't pass it on down. Maybe he was told not to pass it on down. But there was an inadequacy in my opinion, in that flow of information to the Chiefs.

I was determined, as far as the navy was concerned, that we had two principal considerations. First, we had to ensure that there was a full flow of operational information, from us to the president. Second, that there was to be a firm impediment by the higher authorities of the navy for any direct control or interference by our civilian authorities to our operating forces. We did not want, and I had it pretty well set up to prevent, any intrusion by McNamara or anybody else in the direct operations of any ship or squadron or anything of the sort.

THE CUBAN MISSILE CRISIS

After the president said he knew our views, that this was the decision and it was up to the navy, I came back, and McNamara said—I think it was directly or it may very well have been through General Taylor—that he ordered the navy to take care of the quarantine, and that LeMay would take care of reconnaissance, of the flying. So I went back to the CNO's office and I took the three most reliable senior officers I had—Claude Ricketts, Oley Sharp, and Don Griffin, the Vice CNO, DCNO Operations, and DCNO Plans—and directed one of them to maintain a round-the-clock watch in the CNO's office when I was not there and to maintain continuing liaison with flag plot, where we had our central command and control. I had Capt. Ike Kidd, my executive assistant, outside, and one of these three senior officers always in my office. In addition John McCone, director of the CIA, had asked if he could have a liaison man in my office, and I said, "Certainly." It was amusing because he looked at things that we would furnish to him, then pass them to CIA, and CIA would pass them to the White House situation room as if they had originated in CIA.

It was apparent to me, and this is a lesson I had from the operation, that to control the naval operation through flag plot up in the Navy Department section was not a satisfactory way of handling it. It would have been better to have those things all handled by the JCS command post in the JCS area. This means, of course, in the case of the navy, we would have to have better, more highly qualified officers available to stand watch in the Joint Chiefs of Staff command post area than we would have liked to have provided down there otherwise. Also, it might be a little further removed from the superior naval advice and direction, which we could get by having these three experienced senior officers up in the CNO's office. However, we could probably work that out by having them stand watch down in the JCS command post area, if necessary.

We in the Chiefs, particularly LeMay and myself, would like to have had more reconnaissance than was permitted, particularly low-level reconnaissance. This was restrained by the White House and every flight had to be justified to a great degree. They were trying to exercise too much detailed control. I might say that the Joint Chiefs of Staff were all aware of the magnitude of the whole situation. I didn't hear of any one of the senior military people trying to move their families out of Washington, which did happen in the case of some of the civilians who were involved. I think they took it in a rather professional way.

I sent out a directive to make sure that there were qualified Russian-language officers on each ship involved in the quarantine, in case there had to be interrogation, and I made available through the Bureau of Naval Personnel to Adm. Robert Dennison, commander in chief, U.S. Atlantic

Fleet, in Norfolk, Russian-language people, for example from the Naval Academy. Dennison said he'd get them on there, and that was enough for me.

But McNamara wanted me to get into every detail; he wanted me to interrogate each ship as to whether language officers were actually on board. This is an overpreoccupation with detail involvement that I don't think the civilian authorities should get into. I think when you get into a situation of this sort, they can handle the policy level up above; that's their responsibility and their prerogative. But I just resent the involvement of the secretary of defense, and his lower-level civilian staff officers getting involved in military affairs. It's lack of an appreciation of what prevails in the military, lack of custom to command, the system. You would not have had that with some of the people we've had in the Defense Department. But these people had been burned in the Bay of Pigs, and I guess they were so sensitive about the whole thing that it governed or stimulated their lack of experience.

The Chiefs became aware, of course, of various types of negotiations that were being made, but not to the extent of the conversations and the letters that the president had sent to Khrushchev. I've heard that toward the latter part of these negotiations—as a matter of fact, I know this is true—Kuznetsov, who was the deputy foreign minister, approached John McCloy in New York and said, "What does your government want?" And McCloy said we wanted those offensive weapons out of Cuba, and Kuznetsov said something to the effect, "Well, they will be out, but never again will the Soviet Union be caught in the position of strategic inferiority."

It's my belief that at some point after this crisis was over, there was a meeting of the Politburo; whether in December 1962 or in January 1963, sometime in there, they made a decision greatly to expand their military forces, their strategic forces, to extend military aid to Communist countries or countries that they wished to influence in Southeast Asia and in the Middle East. I think that subsequent events have brought out that suspicion that I had at the time, that this produced a major decision on their part, which has led to the great buildup of Soviet military strength. Because, in fact, in and around Cuba, they had challenged the United States in a location where the United States unquestionably had overwhelming superiority, not only in Cuba and the Caribbean but also in the Western Atlantic. It didn't seem to be a very wise action on their part.

Also, I had taken a particular determination that we were not going to let any Soviet submarines get in and start operating out of bases in Cuba. In fact, we concentrated our whole area antisubmarine coverage to the point that every Soviet submarine in the Western Atlantic was made to surface at least once, or several times in some instances. I had excellent cooperation

A U.S. Navy helicopter hovers over a surfaced Soviet submarine during quarantine operations in the Cuban missile crisis, 11 September 1962.

from Admiral Dennison in that regard, and I did follow very intensely our successes in that respect. We were 100 percent successful. At that time, of course, we were dealing only with conventionally powered submarines and no nuclear-powered submarines in the Western Atlantic. If they'd had nuclear-powered submarines, the problem would have been more difficult. Also, as I mentioned before, we did have adequate naval forces on active duty ready to cope with any situation in that area. As a matter of precaution we had heightened readiness of our Sixth Fleet in the Mediterranean and our Seventh Fleet in the Western Pacific. I had no real apprehension that, due to inadequate forces, the navy could not cope with any eventuality that might evolve, except, of course, if the Russians had elected to go to all-out nuclear war, which would have been a most foolish thing at that particular time because we had such a preponderance of nuclear forces.

One incident occurred. We knew where one of these particular submarines was located. We had that information from the most highly classified intelligence the navy had at the time. We were very anxious to preserve that intelligence, and very few people knew about this type of intelligence. We had a destroyer sitting on top of this submarine. One evening, McNamara, Gilpatric, and an entourage of his press people came down to flag plot and, in the course of their interrogations, they asked why that destroyer was out of line. I sort of tried to pass it off because not only were there some of McNamara's people there who were not cleared for this information, but some of my own watch officers were not cleared for it in the general area of flag plot. After some discussion, I said to McNamara—he kept pressing me—"Come inside," and I took him into a little inner sanctuary where only the people who had clearance for that particular type of classified information were permitted and I explained the whole thing to him and to his satisfaction, as well. He left, and we walked down the corridor, and I said, "Well, Mr. Secretary, you go back to your office and I'll go to mine and we'll take care of things," or words to that effect, which apparently was the wrong thing to say to somebody of McNamara's personality.

I heard nothing of that until after the TFX affair, then the story was leaked to the press through his own public information people that I had insulted him by making this remark over the incident in flag plot. And it's still there, they're still correcting it in most stories that come out about the Cuban missile crisis. He took offense at that kind of statement because he's just that type of man. Maybe I was provoking but it certainly didn't register in my mind at the time and was not so intended.

There was another thing that was reported. We have standardized tactical publications for almost every conceivable type of naval operation. They're worked on and revised continuously and they're called "tactical doctrine publications." A commanding officer has those on board ship; it's his doctrine, and he has to follow it, and McNamara was getting into the instructions that these people had. I said, "They have these things, they've had them for years in the doctrine publications that they have as a basis to follow." There was no reference on our part to John Paul Jones, but somebody—it was not I and not one of the naval officers there—reportedly said that McNamara said to me or I said to McNamara, "We've had them since John Paul Jones." It was the reverse that had occurred. It didn't bother me at the time at all, but apparently this is another story that's come out, about how he doesn't care what the navy had done in the days of John Paul Jones.

When the settlements were being negotiated to get the weapons out of Cuba, naturally we wanted to have on-site inspections to make sure that they were in fact removed and that there was no repetition of them. But this on-site inspection is, of course, a *bête noire* of the Russians in any context. So they finally worked out that we could photograph the missiles on the ships going out, and we insisted that we would have overflight inspections. They had to be on deck, and, to all intents and purposes, I would say that the Joint Chiefs were satisfied that the missiles were taken out and that the IL-28s were also flown out. But they have never really implemented the idea that a prohibition could be applied to Castro infiltrating or trying to influence actions in other countries of Latin America, which I think is a very sad thing, and the fact is, in the early 1980s, the Russians still had about three thousand troops in there.

It was clear that we did have a great tactical superiority, we had a strong strategic position, and we won, but we didn't capitalize on what we had. I think that was a sad aspect of it, although, as I've always said publicly, what the president of the United States decided to do was in keeping with the character and the moral posture of the United States. It still, in my opinion, was not the best geopolitical decision that he could have obtained if he had just put more pressure on.

Unquestionably, if we had gone in with an invasion, we would have been successful. It could have been rather bloody but I would say with a relatively low degree of casualties on the part of the American forces. I think the Cuban people would have immediately rallied to our support and I think we could have installed a good government in Cuba. I think with the proper warnings to the Russians and care on our part that there would not have been a military confrontation between the Russian troops and ours, because there were relatively few Russians there. I don't believe under any circumstances they would have fired those offensive weapons against the United States or it would have been nuclear warfare. But, of course, in anything you undertake there's always a risk, with the risk in this particular case being the continued control of Cuba and the expansion of communism throughout Latin America. In any event, that crisis was resolved apparently to the satisfaction of the president and, after all, he's the commander in chief.

I think the military people just showed that they were professional. Obviously, there were some lessons learned, one of which was how we could improve command and control of the military in a more efficient way and, at the same time, prevent the civilians from getting into something they're not qualified to handle. Our intelligence support by our own intelligence people in the DIA and Naval Intelligence was excellent. We gave everything we had over to the CIA. I think that perhaps the flow from the CIA to us could have been improved. I think we lacked in some measure a flow of information on the deliberations at the special group, from the White House down.

This experience also indicated that there were serious weaknesses in having enough commercial shipping, if we had become involved in the necessity for major logistic support, even in Cuba, as close to Florida as she is. If this type of problem occurred thousands of miles away from the United

States, it would be more aggravated, as would be air cargo lift. We didn't do anything about the development of a real merchant marine because of practical economics.

I think the services did a superb job. Southern Florida was practically sinking under the buildup of military forces there. This must have been a deterrent to the Russians, too. We had to take daily aerial photographs of the airfields in southern Florida, particularly of Key West, to make sure that we weren't having undue concentrations where an accident, much less an enemy attack, could have destroyed a large number of our planes. We had to insist that they have reasonable dispersal. We had our carriers in a high degree of readiness and I would say generally everybody performed exceptionally well. The president was highly pleased. After the operation was over, he made a tour in which he invited us to accompany him. He went to the various stations and thanked the people involved, then we went back to normal, including the usual battles over the budget.

=17=
You Only Go Around Once

Admiral John J. Hyland

John Joseph Hyland was born in Philadelphia, Pennsylvania, on 1 September 1912, son of Capt. John Hyland, USN, and Josephine (Walker) Hyland. He attended Brookline (Massachusetts) High School, prior to entering the U.S. Naval Academy in 1930. Graduated and commissioned ensign on 31 May 1934, he was ordered to Pensacola, Florida, for flight training in June 1936 and on 12 May 1937 was designated naval aviator. He subsequently advanced in rank to admiral, to date from 1 December 1967.

Admiral Hyland served with various fleet squadrons and, at the outbreak of World War II, participated in the defense of the Philippines, engagements in the Netherlands East Indies, and the final retreat to Australia while assigned to a patrol squadron. He later became commander, Air Group 10 attached to the USS Intrepid *(CV-11) and participated in actions against Japanese forces during the final year of the war. Command assignments after World War II included command of the USS* Saratoga *(CVA-60), commander, Airborne Early Warning Wing, Atlantic, commander, Carrier Division 4, and commander, Seventh Fleet in the Western Pacific. On 30 November 1967, he assumed duty as commander in chief, U.S. Pacific Fleet. He was placed on the retirement list effective 1 January 1971.*

Adm. John J. Hyland, commander in chief, U.S. Pacific Fleet, 1967–70.

Admiral Hyland's decorations include the Distinguished Service Medal with Gold Star, the Silver Star, the Distinguished Flying Cross with two Gold Stars, the Air Medal with four Gold Stars, the Presidential Unit Citation, and numerous foreign decorations.

I was commander, Seventh Fleet in the Western Pacific from December 1965 to November 1967. During that time I was stationed on a cruiser; one year it was the *Oklahoma City* (CL-91), and the next year it was the *Providence* (CL-82), both of them are 6-inch gun cruisers, and they changed places. It would have made better sense, perhaps, to keep one of them out there all the time, but about every two years new electronic equipment would be available, and you would want to get that on a flagship. At that time, they wouldn't let the Japanese in the shipyard at Yokosuka handle the latest electronic gear. For that reason, the ship that had been there for two years would go back to the States and get a quick fix, and the other one that had been in the States would come out and relieve them.

The flagship was used while it was there on station in the Tonkin Gulf for gunfire support. But its principal role—it had marvelous electronic gear, communications equipment, on-line cryptographic devices, all of which were very, very good—was that of general supervision. I don't remember any cases, though, where I would be on the flagship and would hear that something had happened and then would realize, "God Almighty, they're not taking advantage of their opportunities," and barge in right away and make them do something different. These people were very competent people; they were officers just a little bit younger than I, and I had known them practically all the time that I'd been in aviation. They always were doing just the right thing.

I would go aboard the carriers quite often and all the ships as often as I could. I'd gotten a bit of advice from Vice Adm. Tom Moorer, who had been ComSeventhFlt a few years before me. He told me that he'd found one of the most effective things to do as a fleet commander was mostly to be seen, to get around and just get people to look at you and perhaps let people listen to you talk for ten or fifteen minutes—just get some idea of what the fellow looked like and how he acted, and who was, in some cases, imposing all these silly rules. I think I've been up and down that sling in a helicopter more times than almost anybody for that period there.

We thought it was an advantage to have a mobile flagship, as opposed to having been shore based, because you felt you were taking part a little bit more. At the same time that the war was going on, there was still the same old peacetime business with countries that were not involved. So the flagship didn't spend all of its time, when it was at sea, down in the Gulf. We would make visits to other places in Japan, but you could do that and still generally keep in touch almost as if you were right there in the Gulf. But we had to do that in addition.

The U.S. Navy had two stations for the carriers that were stationed off the shore of Vietnam in the Gulf of Tonkin. One was Dixie Station, which was the southern one, where a ship newly arrived on station would go and get a little bit more squared away. Those strikes were used principally to support the land fighting that was going on just inshore from where that ship was stationed. Yankee Station, up farther north in the Tonkin Gulf, was the other one. It was from those carriers, usually two of them, sometimes more, that strikes were sent up into North Vietnam. Those were by far the most dangerous, risky flights that anybody had to do in those days.

They divided North Vietnam into areas. Some of them were assigned principally to the navy, and the others to the air force—about half and half.

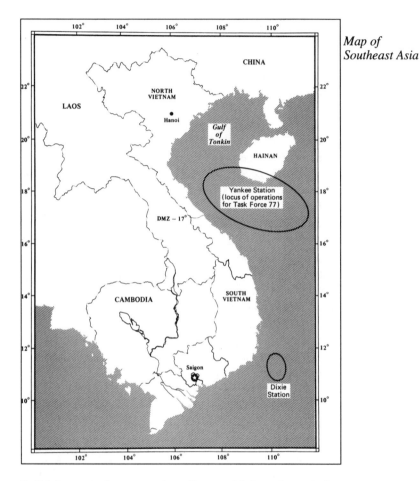

*Map of
Southeast Asia*

It didn't mean that navy air strikes couldn't go into air force areas or vice versa, but normally you did go into your own.

The target lists actually were recommended from out there, sent back through CinCPac and CinCPacFlt, to Washington. It was in Washington that the final evaluation was made and final permission given to go ahead and strike certain targets. It was very, very closely controlled from Washington. In the earlier days, a certain number of targets would be made available to attack. That meant you couldn't attack anything else. That was one of the basic errors in the way the thing was controlled, because just as bad luck would have it, often the targets that you were cleared to strike wouldn't be available on account of weather. We had lousy weather out there, particularly for visual tactical bombing the way the navy did it. But you got your permission to do it from Washington and then went through the whole process of dishing it out to the carriers, and planning on the carriers on how to conduct the strikes. And you would just go ahead and do it as expeditiously as you could.

The navy was really a large support force for the army and for the marines, the way the war was conducted out there—with the one exception of the flights into North Vietnam, and I think the navy did more than its share of that flying. It was in North Vietnam that the pilots began to encounter missile sites, enemy airplanes, and very heavy antiaircraft firing. So it was, by all odds, the toughest flying that anybody was doing out there at the time.

Dixie Station was eventually phased out. There wasn't much for them to do down there, and by the time the war went on for years and years, every ship that came had already been there on a previous deployment. A lot of people were repeaters. For a while there, we had a system where people in the Pacific Fleet were going out again and again and again; nobody in the Atlantic was taking part at all. This was one of the errors in the Korean War earlier; there were quite a few fellows that had four or five deployments in Korea, taking those risks. We had the rest of them in the Atlantic Fleet who were making cruises to the Mediterranean and living it up on the Riviera.

We had every carrier in the Atlantic sent around to Vietnam to do a stint. Of course, it's awfully good for a ship to get it in shape and get it so it is honestly an operating military machine. The best thing in the world is to get them into a situation where they are really conducting flights with live ordnance. So as the thing went on and on and on, all the carriers in the navy took equal part in the process, and that was very, very good for the navy, even though we weren't allowed to win it. It did a great deal for the operating abilities of all those ships and the other ships that were with them.

Rolling Thunder was a code name that was given for the air strikes that were sent up to North Vietnam. My role in Rolling Thunder as the fleet commander was really hardly more than supervisory. We had under the fleet commander the carrier division commanders, the task force commanders. An aviation admiral would be on one of the carriers, and he would be in charge of the entire operation as CTF-77. He would parcel these target assignments out to the carriers that were available there. Then they in turn on board ship would actually plan and run the strikes. But the fleet commander just did his very best to keep in touch with it and know what was going on almost to the hour.

The overall value of trying to interdict the flow of supplies from north to south, in the final analysis, was ineffective, because the enemy were able to reinforce. Now, how much more difficult it was for them, I never could say. It must have been somewhat more difficult, because we did have lots of photographic evidence of bomb craters and destroyed vehicles and train tracks that were ripped up, bridges that were knocked down. But the enemy just seemed to have the old oriental patience, just kept at it. Maybe it was much more difficult for them to get supplies down from north to south, but they did it. They never seemed to lack supplies.

I think it was an effort that had to be made, because there is no question about it; they were infiltrating men and material. We had to try to stop it. There were various schemes to try to do that, but we were never allowed to really carry them out fully. Some people would say, "Well, let's just take one system, say, railroads. Just beat the hell out of the railroads; make them inoperable." Well, the enemy would then just turn to the roads. Somebody else would say, "Well, they can't be doing very much relatively on the railroads, because they don't go every place, but trucks can go every place. Let's go after the trucks." Again, you would be sort of frustrated, because you just couldn't do it. The big crater that you made right in the middle of the road could be fixed very quickly; there was no shortage of manpower. I'd say we did our best, but it was pretty clear that it wasn't nearly good enough to cut off the resupply.

With regard to my command relationships, I was in contact with Adm. Roy Johnson, CinCPacFlt, more or less daily. With the advent of these marvelous communications, instead of writing a message that was a model of terseness and few words, we wrote letters via the radio. At the same time all of this was going on, we were a little worried about naval aviation being absorbed by the air force. Gen. William Westmoreland was the commander down in Saigon—ComUSMACV. I think it's the army point of view there ought to be a single commander, he ought to run the show, and everything ought to be under him—the air force as the air arm of his command—and the navy and the marine air ought to be absorbed into whoever is going to run the air campaign. Of course, from the air force point of view, it's obvious that central authority is the air force.

We could see that, again, we'd have a situation very similar to the one in Korea, where they did have control of all the air strikes, and what happened was they did their best to give the navy the poorest and least important strikes and take the most profitable ones for themselves. So the navy has always been worried about getting itself under somebody else's control and not be able to operate the navy to its best effectiveness. General Westmoreland wanted to have control of everything—for instance, our amphibious operations. The marines and the navy were terrified that might come to pass. So we'd always assign a flag officer to supervise and run these amphibious assaults, but they would be done under ComSeventhFlt. Westmoreland thought they should be done under him.

I can remember one time having a heated argument with him about this. In fact, I told him, "I'll put the most experienced amphibious officer we have in the fleet in charge of this operation, and you won't have to worry about it

On board the USS Kitty Hawk *(CVA-63), 20 April 1967* (left to right), *Vice Adm. John J. Hyland, commander, U.S. Seventh Fleet, Adm. Roy Johnson, commander in chief, U.S. Pacific Fleet, and Rear Adm. David C. Richardson, commander, Task Force 77, discuss the strike on the Haiphong thermal power plant.*

at all. You just tell him what you want them to try to accomplish, and they will accomplish it." He wanted to have a naval officer on his staff run the show and have all these units put under him. Of course, the navy felt that once you got them under him, he'd never release them. It's one of these old political battles, but they have their own importance to the navy and the marines.

I can remember talking to Admiral Johnson on a daily basis about this. My instructions were, "Don't let the Seventh Fleet come under the operational control of MACV in Saigon." They considered it had just the most dangerous potential for the future. I couldn't do anything but go along with that, and it wasn't difficult, because I happened to agree with it also. But that caused some friction. I went to Saigon quite often, and Westmoreland was very, very cordial to me. He was quite a fine gentleman. He always put me up right in his quarters, in very nice comfort. He invited me to the daily briefings they had and to the daily press conferences. I was very impressed with him. He certainly put on a first-class college try to do the best he could.

We didn't have any success in getting more reasonable rules of engagement while I was around. It was a constant, nagging problem; it never was established the way it should have been. It always was very closely controlled. I think they were afraid of a whole lot of things; they were obviously terrified of the war expanding, in particular, I think—although this was never brought out very clearly—that China might come into the war. They didn't want to have any incidents that could be avoided.

We had rules of engagement that were just completely ridiculous and terrible. I thought it was one of the great sins to ask those pilots to try to do anything under the rules of engagement that were then in force. Very generally, they said you can take retaliatory action against any threat. If you see an enemy airplane and he is trying to shoot you down, you can assume that he wants to do you harm, and you can try to do something to him—things as silly as that.

You were not allowed, necessarily, to attack a missile site that wasn't an approved target. If a pilot saw one on the ground, he wasn't supposed to strike it unless it did something against him. Well, by the time it's doing something against him, it's almost too late for him to do anything. So, of course, in a case like that, the instant a pilot, group leader, or anyone saw a missile site on the ground, that should be grounds for attacking it right then and there. There's no sense in waiting to see whether it's going to fire, and if it does, then you can fire back. The rules of engagement generally constrained pilots from taking sensible defensive actions.

Those rules came essentially from our old friend McNamara and the "whiz kids." They were trying to limit the war. They were playing a new game, the handling of a limited war. To them apparently it was a very fascinating thing to try this and try that. Of course, our people in our airplanes were just pawns in this game that these men in Washington found so fascinating. It never should have been that way. I remember many times on board carriers I'd go down and talk to a whole bunch of pilots assembled and try to tell them how well they were doing, and how we were trying to get the whole thing squared away so all these lucrative targets would be available whenever we could get at them.

There would always be questions about rules of engagement. One time one youngster who was kind of indignant stood up and asked me quite sarcastically, "Admiral, if I get a couple of bullet holes right through my body, can I assume that I can take retaliatory action?" I didn't shut him up abruptly as a senior officer, and it didn't offend me; I could understand his motives. So I just did my best in cases like that to try to tell them we were just doing the best we could to get those rules changed so they made a little bit better sense, particularly to the people who were taking these very risky flights.

But, on the other hand, the way we do things in the United States, we follow orders. And if the orders tell you that you can't do this, even though it might make military sense, really you can't do it, and you can't assume that you can do it. I've often felt that there must have been many cases where pilots and flight leaders took things into their own hands in situations and went ahead and did something that made sense at the moment. Perhaps they never reported it; perhaps in the way they did report it would work out that they were within the rules, or pretty close within them.

We had airplanes that had missiles on board that were mainly good for a head-to-head encounter. But the rule was you couldn't fire on somebody until you visually identified him. They were afraid that something would appear on the airplane radar, and it would be a PanAm airliner, or any kind of foreign airplane that you didn't have a reason to shoot at. By mistake you might knock it down, so you couldn't fire until you saw it and identified it. This eliminated the use of the Sparrow missile system that would work and was good. So that's the reason, of course, why all the kills were with Sidewinders, where you get in behind the guy after you've seen him. You either shoot him down with guns, or you fire a Sidewinder at him and knock him down that way.

We were never able to use the best system that some of those airplanes had. There wouldn't be any doubt in ninety-nine cases out of a hundred that whatever was showing up on the radar was an enemy airplane; it would have to be. They didn't want to take that one-in-a-hundred chance that it might not be, that some dummy had gotten into the scene by mistake and there he was, and you'd have an international incident on your hands.

I can remember one specific item where I was so disappointed. We had a youngster who was just flying an airplane from Cubi Point in the Philippines back to one of the carriers. He was flying an A-1 Skyraider. How the dickens this youngster could possibly get so far off, I don't know, but he wandered off. You think of somebody heading west toward the Tonkin Gulf from Cubi Point, but he headed off way to the north and was jumped on by airplanes flying out of Hainan Island and was knocked down. Whenever we had a downed pilot, I promised when I got to be the fleet commander, and repeated it later on when I was CinCPacFlt, that whenever somebody went down, we would make a maximum effort to get them back. We'd do anything that was possible to rescue them it if could be done, and they could depend on that.

Well, in this case, here's this youngster down near Hainan; nobody knew exactly where. I was CinCPacFlt when this happened. Adm. Ulysses S.

Grant Sharp was CinCPac, my superior. We had a heated talk over the telephone. He said, "You cannot go after him!" I said, "We have to go after him. We must. We think he's there." He said, "I'm sorry. They will not authorize any flights in there where you might have a go with the Chinese, and before you know it they would have an excuse to get into it."

We couldn't do a thing about it; it was so distressing. I just kept hoping the ASW carrier, which we had there and which did most of the rescue work, because they had the helicopters—I was just hoping that they would somehow or other learn about it and would do it anyway before anybody had made it clear that this rescue thing of going into their territory was not going to be allowed. But they didn't; as far as I know, we never heard again whatever happened to that young fellow.

The pilots later on would ask me questions about that. "Was anything done about him?" I'd have to be sort of shamefaced and say, "I'm sorry, but, no, nothing was done. We were not allowed to do anything." And they wouldn't like to hear that, because they'd been promised that no matter what happened, we would get them if we could. I think we fulfilled that promise in almost every single case. This was a unique exception.

It was very distressing to see these youngsters lost, and in all cases they were top-notch people. Gosh, we had squadron commanders knocked down. We had really good people, really experienced people, but there was no avoiding that. It was not a declared war, but we were at war, and that's your job. If you're that age, you're supposed to go and use your airplane; you just have to do it. We knew there would be losses. I'm sure there must have been apprehensions about going.

The continuing loss of pilots that became prisoners or were killed was terribly regrettable, of course. At the same time, though, all along we thought that the powers that be were getting the news, that they were going to allow us to be more aggressive, and to hit the enemy harder. We felt that if we were allowed to do what should be done, we could virtually silence those missile batteries. We could devastate the areas where the antiaircraft guns were. Of course, the enemy had the handy little scheme in many, many cases of putting antiaircraft guns right in the middle of built-up civilian areas. We were always worried about extraneous damage that was necessary in trying to hit an antiaircraft site that was right in the middle of town. You're going to damage the town. Again, you're going to get women and children and nuns and hospitals along with it. Nobody wanted to do that. But the officials in Washington were so concerned about collateral civilian damage that everything was constrained. But we felt that if we could really go at the North Vietnamese, we could pretty much silence their defenses.

There was some information from time to time that they were running low on missiles. That was believable, because it's more difficult to make a missile than to make a bullet. We thought we could silence them, and the losses therefore would be fewer. We would be much more free to go at them. If you were not going to be fired on by missiles, you could stay over a target area indefinitely. Their guns are not good enough to get at you at higher altitudes. In other words, you could stay there and look down and have all sorts of opportunities for repeated attacks and that sort of business which would be much more effective. But when you are worried about the defenses the way we were, it mainly is a one-pass attack, and then away you go and come back again a few hours later at best.

With respect to the air power that was used, the marines have what they call their air-ground combat team. They have ground forces, and that would include all the weapons and the tanks, and whatever it is you need. They also were accompanied by a certain amount of air power that is the right amount of air power, the necessary amount of air power for that unit. So here the marines are installed on the beach, and they have got relatively a hell of a lot more air support for their operations than the army had, because the air force is supposed to supply that for the army. The air force for years never took it very seriously. They put all their effort into building up the Strategic Air Command, and they didn't give a damn about tactical air. They began to feel differently as time went on. But here you had a situation where, relatively speaking, the air support available to the marines was far bigger and better per unit than it was for the army down south. Westmoreland, I think understandably, felt, "Well, we should redistribute these assets so that everybody has the same amount." And it would mean that nobody had enough, so the marines would resist that.

We came to an agreement that the marines would take care of their own support. Admittedly, there would be some support available left over from what the marines needed, and that would be made available to other people. That's the way it was done, but it's not really a very workable arrangement, though. We all felt the marines ought not be penalized for having planned properly and equipped themselves properly. They come to war and now somebody wants to take away some of what they have, because the other guy either didn't or couldn't arrange to have himself supported in the same degree. So we think we do it better than other people, and I think there are many, many things to point to to show that. But those are interservice political difficulties; I guess they are always going to arise. It's only in a case like World War II, where everybody had everything he wanted and more, that you're not going to worry about it. Then, I suppose, they will figure out something else to have difficulty with. Those were the things that were involved when I was down there.

The quality of the logistics support of the Seventh Fleet while I was in command was top-notch. We never had any problem except that here and there were shortages of various kinds of ammunition. We never had enough of some of the smart weapons that you might think you could use. Logistics support depended greatly on that base at Subic Bay. Mail was delivered to the kids all over the fleet, practically on a daily basis. Carrier-on-board delivery (COD) airplanes came out to the task force there every day and all day long. Things like that were just first class. I think we had it certainly much better than other people on the beach, although their logistics support was pretty darned good, I think.

I forget who said it, maybe Napoleon: "Give me ten yards of ribbon and I'll win the war," or words to that effect, and I found out it might still be very true. Everybody who's involved in combat operations normally does it willingly because it's his duty. But if he finds out that someone else has been rewarded with a medal—he actually sees this guy walking around with a ribbon on him and he doesn't have one and knows that he's done the same things substantially—he wants one, too. And he's very indignant and mortified that this isn't done very fairly. So one of the functions of a very senior guy is to try to make sure that these things are given out equitably, that someone doesn't just happen to have lots of medals and some other fellow has none, and they've both substantially contributed the same thing.

One thing I've discovered—I was greatly amazed about it—was on two big-deck carriers that had been out there in the Tonkin Gulf. They had been there just about the same time; they had made just about the same number of difficult flights up into North Vietnam. They'd been there the same length of time and so on, and the skipper of one of those ships had out-recommended the other skipper by a ratio of six to one. The old hard-nosed guy said, "What the hell do they need medals for? They're just making flights. It's their duty, isn't it? Unless you can say that this fellow is actually the single person responsible for knocking that power plant out, why should any of them get anything?"

The other guy said, "They're going up there and risking their lives. They are doing the best they can; they are doing damage to the enemy. There is no way to separate them, and for the same exposure to dangers and all that business, they ought to be equally rewarded, and I think every one of them ought to be rewarded to the maximum extent." That's what he was doing. So, boy, we had quite a time. I think I was able to fix it so the thing got changed so it was equitable.

I remember late in my time one air wing commander wrote me a letter in which he claimed his outfit wasn't treated fairly. He was angry in that letter. He made an excellent case that his people had not been give a fair shake. He was still seething over the number of awards that his outfit had gotten. He wasn't thinking of himself. My gosh, despite all our efforts it turned out this guy was right. We'd just missed it somehow or other. He did have a bunch of people that had missed out on getting decorations. So we just adjusted it after the fact, and they got some. I'm not sure he was ever entirely satisfied with the adjustment, but it was a major one. I sure was pleased about that. But that's one thing that you really do have to be careful about. I do think that the awards, medals, and so on should have some integrity. It's bad to prostitute that very much. I tried to tell those people, all of them, at meetings, "We have to be sure about the thing being fair, but for goodness' sake let's not go overboard and ruin the whole system so that the only thing a really good guy could get that would make him stand out from others would be to get him a Medal of Honor, and there's not many cases where you can do that." Usually the guy is dead, and it's done that way.

The pilots who were flying patrol planes out there were, in most cases, making just area patrols that were absolutely no different at all from peacetime employment. They were unhappy as they could be because they could not qualify for a Distinguished Flying Cross and above, simply because they weren't involved in combat operations. I think that many of them are still disappointed about the fact. I made the decision that if any of them did get involved in combat by any happenstance, we would certainly consider that. But for simply making a flight out of Cubi Point and counting ships out in the South China Sea, that really wasn't any different in those days from doing the same kind of thing in the Mediterranean. It wasn't worthy of medals. I think there were one or two new medals that were developed. They were supposed to fill in the gaps between giving someone a very high decoration and giving him one that had been so liberally awarded that it didn't amount to a damn.

I've always counseled people to go as far as they can in awarding people. Now it really is only a psychological thing. When someone does an absolutely excellent job, write him a letter of commendation or pin a medal on him, and it will motivate him to just break his back from then on. On the other hand, you should not pin medals on people for just coming to school regularly.

It is appropriate to mention the Positive Identification Radar Advisory Zone (PIRAZ) ships, the DLGs and so forth that had Navy Tactical Data System (NTDS). A PIRAZ ship had excellent radar gear, and it kept track of people who came in and who departed combat areas. They knew when the attacks were going to be made and where. They could be very helpful sometimes in assisting in the rescue of downed pilots, not only navy but anybody else who happened to be up there. I think it was an excellent use for the ship. It was the most sophisticated vessel I guess I've ever been on that was not an aircraft carrier. I've often wondered about how they would function if they had battle damage. How do these ships that really pretty much absolutely depend on computers, how are they going to perform if they have damage? There always is some manual way to do these things if it's necessary, but the way things go, you never have time to do it. You have to spend enough time to get to know how to use the things via computers. There's never really any time to become good at doing it manually. I often wonder, "What the dickens are we getting in for?"

As the system was being developed, they were managing an area that was a navy route package. The air force at first resisted having to report to the navy when they were coming into that package. They later found out that it was adding to their safety very much to do it, and they were quite willing to do it. But, at first, they had the same hang-ups. "If we're going to have to report to this guy and see if it's okay to come in, isn't that in a sense the navy beginning to monitor what we are doing?" They didn't like that. We felt the same way; this is penny-ante stuff really. You can never get completely away from it.

I think the tempo of operations was very high. Everybody involved had a six- or seven-month deployment. In other words, every youngster in the Pacific Fleet could look forward to every year being away from home for about six months. I think a lot of people, when you mention that to them, their reaction is, "What the hell of it? They are sailors, aren't they? That's what they're supposed to do. They're supposed to go to sea." A lot of people think sailors are supposed to go to sea and stay there. Well, Adm. Hyman Rickover thought you ought to go to sea with a nuclear ship and stay there; that's what it was built for. But in peacetime and in an undeclared war, that is pretty tough on people. So if we'd had a lot more ships, we wouldn't have had such long deployments. But we had plenty of ships to do the job.

The Seventh Fleet was very large in my day, compared to what it is now. The whole navy had more ships. I think we have always been reluctant to turn in old ships. We never want to turn anything in until we have something new to replace it that's proven itself. Of course, this develops a conflict of interest. I think we are guilty of holding on to very old ships that aren't doing a heck of a lot for us, and they do require crews. We keep them, because we don't have anything else in the offing, and we don't like to lose anything. Nobody volunteers to give up stuff in the military.

We opposed the bombing halts, but we accepted them, of course. I never felt they were going to do any good. Apparently it was a political gesture to indicate our willingness to negotiate. That's supported by a whole lot of stuff that I've read since then, but it simply was not effective. It never achieved any particular purpose. I'm still impressed by the fact that when they were being so difficult toward the end and the president ordered heavy strikes against Hanoi and Haiphong and the mining of Haiphong, the North Vietnamese promptly agreed to negotiate. Normally, during bombing halts we'd just stay on station and talk about it. I don't think any of us in the navy had any faith that they were going to accomplish anything. Naturally, when the war had gone on that long and you have a cease-fire and it's for this purpose of trying to bring an end to the thing, nobody's going to regret the end of a war, for heaven's sake. But it's just that we had no particular confidence that it was going to bring an end to the war unless we were going to end it entirely on the other guy's terms. It certainly was overall the most unhappy experience in our history.

I was still commander, Seventh Fleet at the time of the USS *Forrestal* (CVA-59) fire on 29 July 1967. By accident someone fired a Sidewinder right on the deck, and it went across the flight deck and hit another airplane and set off a fire that turned out to be a conflagration. I was up in Yokosuka when that fire started. She had just come on station down there. It was one of those things; we'd done it to ourselves. I flew down right away. Rear Adm. Harvey P. Lanham was the division commander riding on the *Forrestal*. We inspected the ship; God, the damage was just horrible. The thing was mangled, so we had to send it back to Norfolk to get it fixed. They were stoutly maintaining that they could fix it right at Subic Bay, which was asinine. The skipper of the *Forrestal*—and he was a hell of a good guy—was just absolutely desperate to figure out some way that they could stay on station and carry out some part in the war. They just couldn't do it.

There were two other carrier fires during the period I was out there. The *Oriskany* (CVA-34) had one in October 1966, and the *Enterprise* (CVAN-65) had one in January 1969. The one on the *Oriskany* happened when she was out there in the Tonkin Gulf. It was another one of these cases where we'd done it to ourselves; it was not enemy action. It's just that you do have that potential there. We think that we know more all the time and we can handle it, but it's still, to me, one of the real counts against you—you've got all that stuff in one basket, and you now have begun to set up a record of doing it to yourself. The air force doesn't have to do it to you. You're just doing it to yourself, and it's highly likely that you'll keep on doing that. I don't like to believe that stuff, and I really don't. I do say that we have to exert a lot of care and attention all the time. Anytime you're flying airplanes off those ships, in particular when there's live ordnance out and exposed, you just have to be terribly careful. You just can't permit any of these misadventures that we sometimes have.

The *Enterprise* was operating off Hawaii on her way out. It happened on board her; she wasn't quite so badly damaged. We did lose quite a few people there and a lot of airplanes. But she wasn't all that badly damaged and we were able to fix her right there in Pearl Harbor. Of course, they were absolutely overjoyed and delighted to be able to be given that job. It is an expensive thing to send out very heavy armor plate that goes on the flight deck, where it's been mangled and so on by the exploding bombs—get it flown out there and put it on the ship. They did it in record time. I can remember how the whole shipyard never has had quite enough work to really keep it going in a flourishing way. Just one big job like that means a great deal to a relatively small yard. They were very proud to have fixed the *Enterprise*, and pretty soon out she went.

We had many outstanding carrier division commanders out there. Rear Adm. J. D. Ramage was a particularly good division commander. He was a very colorful individual and a very knowledgeable guy; he knew what airplanes could do. One time it was proposed by the air force that we just put a whole flock of airplanes in the air. They wanted to make a big raid over Hanoi or Haiphong, and they were going to do it above the clouds. We were just going to tail in behind these big airplanes, and whenever they pulled the string, our guys could see the bombs falling from the ones up ahead, and they would just press the button then and drop their weapons. Their idea was sort of to bomb Hanoi from above the clouds, no sight of any target, fully at risk from the missiles and all that. It was the most asinine mission that anyone could dream up, but the air force can and does dream up missions like that, ones that we would never even think of. You know, you tell a pilot that he was going to do this, and he would say, "What?!"

So Jig Ramage refused to do it, and he got some mild criticism. Adm. Elmo Zumwalt, CNO, didn't like that. He thought Jig was being uncooperative with the air force. To be the CNO and not understand that any better than he did to me is unbelievable, but Jig never got beyond two stars. He would have been a fine AirLant or AirPac commander. He was senior, just right for it, but as long as Admiral Zumwalt was in charge there, it just never was going to happen. So Jig finished up as the guy down in the Caribbean. He was a damn good officer, very much unafraid of seniors and very, very biting about his opinions on ridiculous missions like the one I just described.

Looking back on the job of CinCPacFlt, I don't think it really was any more demanding than being commander, Seventh Fleet. It was just as demanding, though; both of them had the characteristic that you couldn't get away from them. You were on duty twenty-four hours a day; you were always on call. Nobody really objects to that, and nobody would want it changed. You grouse about it all the time; "I can't get away from that job."

You don't want to give up the job because it's so demanding; I can tell you that. The biggest frustration was the growing knowledge that I didn't have very much to say about anything. Awfully smart people are planning the deployments. Out of this many ships, how do you keep this many on station? There is no way I could look at that and say, "My God, there's a better way to do that. We can easily do it some other way." Hell no, there isn't any other way than the one that's presented. So you're mostly in the business of looking at things carefully, but approving them because they are so well done. If anybody did come to you with a scheme to handle this or that and it's so loose and so full of holes that even you at first glance can do better, I think you can start thinking of getting rid of this guy, because he ought to be able to do better than that when that's his sole job.

But I was one of the luckiest guys, I guess, that ever came along in the navy. If I could do it again, I'd do it a lot better. But you only go around once.

=18=
The *Enterprise* in WestPac

Vice Admiral Kent L. Lee

In January 1967 I was notified that I would receive orders to be commanding officer of the USS *Enterprise* (CVAN-65), the first nuclear-powered aircraft carrier. I was to spend February and March over in Admiral Rickover's nuclear power program and participate in a refresher course for commanding officers. All prospective commanding officers of submarines were brought into Rickover's office for two months, run through a refresher course in all phases of nuclear power, and given a final examination. I was to join one of these classes.

I think the nuclear power training that I had completed in 1963 was invaluable for me for both the *Alamo* (LSD-33), my deep draft command, and for the *Enterprise*. In my previous training I had a lot of technical training, with a minimum of exposure to naval engineering. In terms of the *Enterprise*, I never, obviously, operated the engines, but I think it's very important for a commanding officer to be able to talk the same language his engineers talk. In my previous nuclear training, I had qualified as an engineering officer of the watch, so I knew what their problems were and I knew what they were talking about. I spent a total of fourteen months in Rickover's office and then an additional two months on board the *Enterprise* before relieving the commanding officer, Capt. James L. Holloway. I don't think I needed that extensive kind of training. I think five or six months total would have been adequate, but Rickover had a policy that the original training would be one year to the minute.

I finished with Rickover in March and went to the West Coast for various courses, such as fire fighting, damage control, and other subjects useful for prospective commanding officers of carriers. I then headed west, and sometime around the first week or two in May, I got to Subic Bay and took one of those carrier-on-board delivery (COD) transport planes out to the *Enterprise*, which was in the Tonkin Gulf. The plan was that I would spend two months aboard the *Enterprise* as an observer and then relieve Holloway back in Alameda in July. I reported aboard the *Enterprise* feeling that the last thing any ship needs is two captains. It was a very awkward situation, being a prospective captain, wandering around the ship with not enough to do. I tried during that period of time to visit all the departments on the *Enterprise* and to learn all about the air department, the operations department, the RA-5Cs, and the intelligence function. I wanted to get as attuned as I could to the operations in Vietnam.

I toured all the ships in Task Force 77 including the flagship, which Rear Adm. Ralph W. Cousins began to command about this time. I thought Admiral Cousins was just about the ideal man to be commander, Task Force 77—a very balanced, intelligent man. I served with Cousins, in one way or another, two or three times after that. He was always the perfect gentleman, always did the right things at the right time, very loyal, very cordial to his subordinates, a first-rate naval officer in every way. He would have been my candidate to be chief of naval operations when Admiral Zumwalt got the job—not that I had a vote.

After these two months, we returned to the West Coast, arriving around 1 July 1967. I relieved Holloway on 7 July 1967 and, after a very few days, took the *Enterprise* to Hunters Point Naval Shipyard for seven weeks. We had a restricted availability, which is a navy term for getting modifications and repair work done after a long voyage or a long cruise. After Hunters Point, we had refresher training down in San Diego.

The *Forrestal* fire took place during this period, and helped to focus our attention on damage control. I instituted a training program on damage control and fire fighting after we left San Diego. We probably had more training in damage control and fire fighting, all aspects of it from the flight decks to the engine rooms, than any other carrier in the Pacific Fleet. We really worked at it. The performance of our ship in the fire the succeeding year attested to that fact. Everything worked, and we fought it the way it should have been fought. We developed, I thought, a fairly good damage control

The USS Enterprise, *the first nuclear-powered aircraft carrier, as she appeared during an early deployment in the Atlantic Ocean in 1962.*

and fire-fighting organization. It's a constant struggle; you really have to maintain that level of training at all times.

The *Enterprise* was a dream to handle. My first experience in handling ships was with those twin-screw YPs at the Naval Academy. I went from that to the *Alamo*, which was a twin-screw, single-rudder ship, but with a flat bottom. You had to be very careful with the *Alamo* when coming alongside another ship, or you'd be sucked in for an alongside collision. The *Alamo* was a very nice-handling ship under most circumstances. However, I found that the *Enterprise* was even easier to handle than the *Alamo*; she was very responsive to power and to rudders; the engines reacted easily to command changes. I could go from full power to stop or from all-stop to full power immediately and not have to worry about boilers. It did take miles to stop the *Enterprise* once you were doing 30 knots, even with all power astern, though.

The *Enterprise* had a very fine engineering plant. Whereas the *Forrestal* class had four propellers and two rudders, the *Enterprise* had four propellers and four rudders, which made it a little more responsive to the rudder.

The *Enterprise* was a little longer and bigger, had a better fineness ratio, and had greater horsepower—280,000 shaft horsepower. We actually had more horsepower but were limited at top speed by torque on the propeller shafts. In the first cruise, we were limited in top speed to 32 or 33 knots because the bottom was a little dirty. In 1968, we went into Bremerton to get the bottom cleaned, and we probably had a top speed of 35 or 36 knots, still limited by the torque on the propeller shafts. Overall, the *Enterprise* was a marvelous ship to handle—a shiphandler's dream.

On 3 January 1968 we loaded our air wing aboard. We had an air wing commander by the name of Paul Peck, who was a very able, aggressive wing commander. The air wing assembled from all over the country. We had two squadrons of F-4s from Miramar; we had an A-6 squadron from Whidbey Island. I think we had A-4s in 1967 and 1968 from Lemoore. We had an RA-5C squadron of six planes from the East Coast. We also had some cats and dogs, such as helicopters, radar planes, and our own COD planes. We had numerous airplanes on board and about fifty-five hundred people when we departed Alameda Naval Air Station. Our eventual destination was to be Vietnam.

After some intermediate stops and a port visit in Sasebo, Japan, we departed with plans to join Task Force 77 off Vietnam. We planned to stop in Subic Bay en route. Shortly thereafter the ship's communication system intercepted a strange message from a ship by the name of *Pueblo* (AER-2), and it was saying something about, "I'm being attacked. Help, help!" At the time, Rear Adm. Horace H. Epes was on board; he was in command of the task group, which was a part of the Seventh Fleet. When we received the *Pueblo* message, we didn't know that there was such a ship as *Pueblo* and we didn't know that the *Pueblo* was in the Sea of Japan. We had no idea what or where the *Pueblo* was. We had no information on the *Pueblo* at all. Another factor that prevented us from immediate action was that the *Enterprise*'s deck was not spotted for launch. We'd been in port and the planes were being worked on. The flight deck was filled with airplanes, and the hangar deck was relatively empty. We had maintenance going on in the hangar deck. All we had were two F-4 Phantoms on the catapults that, in a very short time, could be launched as combat air patrol (CAP). We might have had two backups, but as for A-6s or any other type of aircraft ready to go (except for those two F-4s on the cats), we had none.

Hindsight is always great. It's been suggested that Admiral Epes should have launched on receipt of the message. He could have always recalled those airplanes. That's true, but by the time we waited for clarification on the messages and found out that the *Pueblo* was a U.S. Navy ship in the Sea of

Japan and that it was a trawler-type, communications intelligence ship, it was too late to launch. I would say it took two or three hours to find out these crucial pieces of information. In the meantime, we were getting our flight deck all set to launch. The other odd part of this situation is that the *Pueblo* didn't report to commander, Seventh Fleet. The *Pueblo* reported to commander, Naval Forces Japan, Rear Adm. Frank Johnson, which was a shore-based command. He had operational control over it. Johnson, I believe, reported to CinCPac rather than to CinCPacFlt or Seventh Fleet, so the *Pueblo* was not one of the Seventh Fleet's ships, unfortunately.

How much knowledge Seventh Fleet had of the *Pueblo*, I don't know. But the *Enterprise* had none, which was unfortunate. Our intelligence personnel didn't know and hadn't been informed, either. That was unfortunate, because we should have been informed of everything out there, but we were not. No instructions or orders from higher authority were received to advise us to take any action. We just intercepted this message. Epes could have launched on reception of the message, and in hindsight, I suppose that's what he should have done—decided to launch the two F-4s we had on CAP and vector them up. Our F-4s probably could have reached the *Pueblo*, and the *Enterprise* could have made best speed toward that site. That would have looked good, but by the time it was sorted out—we were in touch with commander, Seventh Fleet and commander, Naval Forces Japan—it was too late. I think it's problematical whether or not we could have done anything anyway. But that, essentially, is what went on in the *Enterprise.*

We were then ordered to head up through Tsushima Strait into the Sea of Japan. There we started low-level flight operations, exercising our pilots and our planes, primarily CAP and the various other planes that needed a little practice. The navy put on a large show of force in the wake of that seizure on 23 January. We steamed around in the Sea of Japan, and additional ships were sent up to join us, from 23 January until maybe 13 or 14 February. In the meantime, Russian ships came down from Vladivostok and other Russian ports in the northern areas of the Sea of Japan. We saw the Russian ships almost every day, because we had gone fairly far north into the Sea of Japan, perhaps opposite Wonsan, the big North Korean port where the *Pueblo* had been seized.

We were moved out of the Sea of Japan in a most unusual manner. One evening I was on the bridge at perhaps 7:30—I would assume this would have been about 7:30 in the morning in Washington. I was told that I had a telephone call in the little tactical plot area behind the navigation bridge. I went back to take the call, and much to my surprise, found the call was from President Lyndon Johnson in Washington. He wanted to know about those fellows from the north; were they giving us a problem? I told him that there didn't seem to be any problem, that we were doing very well. I didn't think there was any great danger. He told me then that what he wanted us to do was to turn south and head out of the Sea of Japan. The duty officer from CinCPac was on the line, and he asked me after the president had hung up if I fully understood what we were to do. I said, "Yes," and that I would immediately turn the formation south and head toward Tsushima Strait.

This communication was not authenticated at the time; there was no way I could be absolutely sure it was the president. Consequently, CinCPac said they would send an operational immediate message to authenticate the president's orders. I felt it would do no harm to turn south immediately. However, we had an admiral on board and, of course, you don't make such dramatic changes without his approval and usually his direction. In those days, admirals and their staffs worked on the strike operations, and the flag captains ran the ships and the formations. This was the division of labor.

Admiral Epes liked to enjoy a movie every night. The instructions were very firm that the admiral was not to be disturbed while he was having his movie, but I thought this message important enough that I go disturb the admiral. I went down to the flag cabin and called the flag lieutenant aside. I explained that I had to see the admiral; this was very important. The flag lieutenant turned a little pale, knowing that this might be a problem, but he agreed that the movie would be stopped and I would see Admiral Epes. When the lights were turned on, I went in to see Admiral Epes and I told him that he wouldn't believe this, but I had just talked to President Johnson on the telephone. The president had ordered us to head south out of the Sea of Japan, and CinCPac said that a confirming message would be sent immediately. Admiral Epes was a little bit taken aback and miffed that orders having to do with his task group came in this fashion. Nevertheless, we headed south out through Tsushima Strait.

On 15 February the *Ranger* (CVA-61) came up to join us, and Admiral Epes was moved to the *Ranger*, which was left in the Sea of Japan area. The *Enterprise* headed south, first to Subic Bay and then out to the Tonkin Gulf and to Task Force 77. We also were joined by a new admiral, Rear Adm. John P. Weinel, better known as "Blackie" Weinel. Admiral Weinel was perhaps the best operator that I worked with during the Vietnam War, and I worked with a number of them. He was very knowledgeable about airplanes and weapons. He was also very well informed about the politics of what we were doing—what was going on in Washington and in the world, and the importance of the various things we were trying to accomplish in Vietnam.

During our time out there, it became popular to hit Hanoi, using A-6 Intruders, which were built for lone night attacks. There were no other planes operating in North Vietnam at night in this fashion. Unfortunately, we learned that if we sent lone A-6s up into Hanoi, these airplanes would be picked off. It wasn't all that difficult for the North Vietnamese with their Russian help to pick off the A-6s. After we had lost several, Admiral Weinel was, I think, very instrumental in bringing about a change in these A-6 attacks in North Vietnam at night. Yes, we could send our A-6s up there, but by and by we wouldn't have any left. It was a very dangerous operation. Instead, a better way to attack North Vietnam would be with a large group of planes in the daytime, rather than sending those lone A-6s at night. Otherwise, it was clear we would lose them all. I daresay we lost four or five.

In all justice to those flight crews, they kept coming back. We had a very professional group of naval aviators for that Vietnam War—I think far better than we deserved. I was tremendously impressed with them in my three tours of duty. I spent two months with Holloway and I had two tours as commanding officer of the *Enterprise*, so I saw three different air wings. The average captain of an aircraft carrier tries to have a very close relationship with the air wing commander and the squadron commanders. I tried to get to know the wing commander and the squadron commanders and have them for dinners in the captain's mess, and discuss with them their problems, so I thought that we had a very good relationship.

I was tremendously impressed with the quality of the air wing commanders, squadron commanders, and executive officers of the various squadrons. In all of the pilots, I thought we had a very high order of courage, professionalism, skill—all the qualities that we wanted in our naval officers. It did take great courage to fly over there and face those missiles day in and day out—hundreds of missions and to take a chance of being a POW in Vietnam. During the time I was over there, I never saw any flinching on the part of our pilots. I came back with nothing but the greatest admiration for the group of young pilots who were doing all this in the sixties and seventies.

The trouble with Vietnam was that the North Vietnamese didn't have any air power to speak of. Their fighters were really not a menace; however, they did have surface-to-air missiles. They could move these around North Vietnam and set one off when a pilot was least expecting it. The surface-to-air missiles, small arms fire, and AAA knocked down the great bulk of the planes we lost. We really had no good defense against surface-to-air missiles. We did have various black boxes in our radar airplanes, which would deceive the surface-to-air missile. These ECM airplanes would jam the radars needed by these surface-to-air missiles. But we had twenty-four-hour operations over Vietnam, day and night, and it was just impossible to have the radar planes everywhere they were needed.

I think it might be worthwhile to describe the operations in Vietnam; they really were almost unbelievable. Each aircraft carrier flew a twelve-hour day. This meant that we normally launched our first flight of the day at noontime or midnight, depending on which twelve hours of the day we were assigned, and continued launching twelve or fifteen planes every ninety minutes. This would go on all afternoon and evening until midnight. We would launch our last flight at 10:30 P.M. and recover this particular flight at midnight. We always commenced the recoveries at midnight. Sometimes it would take us twenty or thirty minutes to get them on board.

After that we would refuel and replenish. The *Enterprise* wouldn't need fuel oil but we needed aviation gasoline or jet fuel, ammunition, and bombs. We were unloading bombs in Vietnam at a rate unheard of before the Vietnam War. These jets could carry big loads. We were sending hundreds of them over there every day. With replenishing occurring after the last recovery each day, the average workday on a ship like the *Enterprise* for those people who didn't have an alternate was sixteen or eighteen hours a day. By an alternate, I mean there was no alternate for the captain, and for most of the crew, they were either on a watch basis or they were on a shift basis. For instance, our aircraft maintenance crews were on a twelve-on and twelve-off shift basis. We had about half our maintenance crew come on at noontime and work through until midnight. The other half would come on at midnight and work through the next day.

The people on the watch, such as the officer of the deck and the engine room and CIC people, were usually on a one-in-three watch basis. This was in addition to their regular work. This went on week in and week out, seven days a week. The weather was marginal—rarely any good. Sometimes we'd have to delay or cancel a launch because of the weather, but not very often. Night and day in terrible weather we'd launch these strikes, weeks on end, months on end. It took a terrible toll on everybody. We would stay on the line for three weeks, a month, and then go into Subic Bay in the Philippines for maybe five days. It was then back to the line again.

Very often, as soon as flight operations and replenishment were over, I would approve the night orders and go to sleep. This was usually between two and three o'clock in the morning, assuming we had that particular shift. We would begin flight operations the next day at noontime. Chances are I would go to bed at two or three in the morning, be awakened once or twice during the night, wake up the next morning between eight and nine, and start in again. Now very often we would have to land aboard CODs and other aircraft. The captain of the carrier is always on the bridge or present when flight operations are going on. This meant I got five or six hours of sleep a night. When we went into Olongapo in the Philippines I'd rest up. Otherwise, it was a twenty-four-hour-a-day job.

I can remember one particular night we had launched a strike into Vietnam. A pilot radioed over Vietnam that he was on fire, and that he was heading for the ship with an escort. We told him to come ahead and could see him coming over the horizon, flaming. When he got to within a few miles of the task force, we told him to eject—this was at night—and leave his airplane and let it crash into the sea. We would pick him up. He ejected safely and landed in the water. Our destroyers picked him up and had him back to the ship within a very short time. This was routine for operations in Tonkin Gulf. Meanwhile, we continued with our launch and recovery.

Night underway replenishments—unreps—weren't very different from day unreps. The ammunition ships, refrigerator ships, stores ships, and oilers were all near by. We went alongside one or more of them almost every night. The idea was that you didn't have time to do them all at once. Instead, one night we had an oiler, the next night an ammunition ship, and the next day we would have a stores or grocery ship. We didn't have to go alongside the stores or refrigerator ships more than once a week or ten days provided we could get a large enough supply. The ammo ships and the oilers usually came every third day; we never liked to get lower than half full of the ammo and jet fuel. We would also load up when we were in Subic Bay.

Prior to an unrep, we would be told the location of the oiler or replenishment ship. We would then try to spot him about ten or eleven at night. After our last launch was completed at 10:30 P.M., but before our last recovery at midnight, we would move toward this replenishment ship so that it would be within a few miles of us. We knew what our recovery course would be, and we would take charge of this replenishment ship and move it out in front, in the direction we would be going to recover aircraft. As soon as we recovered the final aircraft, if we had done our planning properly, the replenishment ship would be within a few miles of us. We would then give it course and speed to take up for replenishment. There's a system of lights for making an approach on a replenishment ship. After doing it every night for months on end, you get pretty good at it. I would come in normally at 27 knots, because that was an easy speed for the *Enterprise* and easily controlled. Perhaps half a mile back I would go down to 20 knots, and then maybe 15 knots as we were approaching the other ship. We would normally replenish at 12 to 15 knots, depending on circumstances. The plan was to get alongside as quickly as possible. After a while, we got very good at it.

Within a few minutes of the recovery of the last aircraft, and while the mechanics were refueling their aircraft from the last flight and getting organized for the next day, the crew on the hangar deck would be getting ready for replenishment. As soon as we went alongside, the line was shot over at two, three, or four replenishment stations. The replenishment ship would be ready for us; everything that we needed or were expecting would be there. By the time the two ships were alongside each other, the produce and merchandise and the oil would be flowing across. We always came along the port side of the replenishment ship; we would be looking down on it from our bridge on the starboard side.

I found out that at night, distances can be very deceiving. I always used radar to approach a replenishment ship until the final approach. Then for the final approach, I would come out to the wing of the bridge and observe the wake of the ship ahead. A wake is absolutely indispensable because night vision is not reliable. You'd use binoculars and floodlights, but your depth perception at night is not very good. Once you get the wake nailed down, and after you've made a few approaches, you can come to within a few feet of where you should be. I also had one of those alidades made. If I wanted to be 120 feet out, I would put that 120 feet on the edge of the wake, come right alongside, and there you were.

The first few times on these underway replenishments, and until I could recognize errors, I made all the approaches. Once I had confidence in my judgment and my own eyes, both night and day, then I would let the executive officer and department heads, including the air wing commander and squadron commanders, come up and take their turn. I let them conn the ship alongside. They liked that and it was a chance to talk to them. It was a challenge, something to talk about in the wardroom. I thought it was very important to them too. Most of them became very proficient at it.

The captain always has a very special relationship with his officers of the deck. You have to have confidence in them, and you have to take them into your confidence. It's most important that they not be afraid of the captain, that if they're in doubt, they know they can call him. I let them deal with the ship just as I did the department heads. I let them make approaches and maneuver the ship. Many of them became very proficient; I had a very fine group of OODs.

There was a pattern of operations for the ship during the air strikes. Each carrier was assigned a point—point alfa, point bravo, point charlie—in the Tonkin Gulf. There were some points up in the north and some points down south. We were expected to operate within about fifty miles of that point, something like that, depending on wind or weather. Replenishment ships and CODs coming out knew generally where to find us. We would be at point charlie, and we could always pick up ships and aircraft on our radars.

Since we were in this relatively confined area, it was predictable where our strikes would be coming from. Admiral Weinel tried to vary patterns to deceive the enemy. We tried never to repeat our strikes. This sort of thing went on every day. Our pilots tried all types of deception: high, low, various routes; every day was different. Weinel was very good at that. We were bombing Vietnam, both North and South Vietnam, twenty-four hours a day. Every day somebody was over there, so it would be hard for them to miss our planes day in and day out. I don't know how many sorties we sent over North and South Vietnam during that long period but it was a tremendous number. It would stagger the imagination—the cost—because the *Enterprise* had four full tours out there alone. I don't know when the first strikes started in Vietnam, and I don't know when the last ones ended, but it seemed to me as though it went on forever. I almost made a career of it, starting with the *Alamo*.

We knew that some losses were going to be inevitable. I think if you're going to be a naval aviator and a naval officer, you have to come to terms with that sort of thing fairly early in life. We had losses in World War II, really very large losses. In one squadron I was in, of about fifty pilots, we lost twenty-two or twenty-three pilots; not all were out of that group of fifty, because as we lost pilots, replacements came in. In the other squadron I was in during World War II, we lost about the same number. The point I'm making is that you come to terms with this, or you become ineffective. You have a nervous breakdown, and people do that. You have to come to terms with this if you're going to be a naval aviator, because you have not only combat losses but accidental losses through the years.

I think most naval aviators come to terms with death. It's not that they don't mourn the loss of their friends and their colleagues. It's not that they become hardened. It's the fact that they have to carry on, which is most important in the military; you can't quit, otherwise, you get clobbered. I think you sort of compartmentalize your mind. You're grieving greatly, but you put it aside and go on with the job at hand. Those people who can effectively do that survive and become very capable combat leaders and officers. Those people who cannot, who become emotional about losses, become very ineffective. Now, that may seem a very hard way to look at it, but I think it's about the way it happens.

While the commanding officer is concentrating so heavily on operational matters, the executive officer has to play a very large role, taking care of all the other things that need to be done on an aircraft carrier. The executive officer is a very key man; all the department heads are key, but the executive officer is the manager of the hotel. The captain is running operations, meeting the dignitaries, and doing all the things that only the captain can do; but the executive officer is the man who has to run the whole machine. He has to keep close track of his department heads, division officers, and men; he has to handle discipline and the chief master-at-arms; he has to handle all the visitors and the wardroom. I think being executive officer of an aircraft carrier is the toughest job in the navy. It's tougher than being captain, I think. That was why I wasn't all that eager to be executive officer of the *Enterprise*, because I had seen many executive officers and knew what the job entailed. I was lucky to have two very fine executive officers—Sam Linder, followed by John Alvis, both of whom became rear admirals. Both were highly educated people, both Naval Academy graduates, both in the top of their class.

Now one of the great treats we had on the *Enterprise* for this particular tour was a trip into Hong Kong. Maybe half to two-thirds of the way through our tour in 1968, the British governor general in Hong Kong decided the *Enterprise* could come into port. In its previous tours it had not been allowed in Hong Kong; it had been restricted to Subic Bay as sort of a home port. But, in this particular case, they decided to let us in. We went into Hong Kong and had about five or six days there. That was a great treat for all hands. But other than that, it was Subic Bay and Olongapo, and Olongapo and Subic Bay, and back to the line. This went on until perhaps the first week in July, when we left for home. I know we arrived home on 18 July, home being Alameda in the San Francisco Bay area. After having left on 3 January, we arrived back on 18 July, which meant we were gone about six and a half months.

Six months later we left Alameda on the morning of 6 January 1969. We had about five days to Pearl Harbor, and then several days in Hawaii while we prepared for our operational readiness inspection (ORI). On the morning of 13 January, we got the *Enterprise* under way from Pearl Harbor to head out to the operating area and begin our ORI. On the morning of 14 January, we had a launch at 7:00 of fully armed airplanes. This was our final examination before another trip to Vietnam with a new air wing and lots of new people. The first launch, about twenty airplanes, went off to attack some targets in the area southwest of Pearl Harbor. There was a wind of 12 or 15 knots, and we were headed downwind at about 10 knots so that the wind was blowing over the flight deck. We were getting ready for our second launch, which was to go at 8:30, and we were preparing to recover the first launch.

At 8:19 I was in my sea cabin, probably doing paperwork, when I heard a very sharp explosion, like the explosion that's set off by underway training group people when they start an exercise. The captain is never supposed to be surprised by these because the underway training groups and ORI people tell him beforehand. Nothing like that was scheduled. The moment I heard this, I immediately went to the bridge. In the very few seconds it took to walk from the sea cabin, which is not more than five steps away, to the wing of the navigation bridge, the whole aft end of the flight deck was enveloped in smoke and flame.

I immediately realized that we had had a very bad accident. Something had set off an explosion on the flight deck, and it had punctured fuel tanks and ignited a holocaust on our after flight deck. The first thing I did was take the conn and turn the ship into the wind and order general quarters immediately. Now there was a problem with ordering general quarters, because when a ship like the *Enterprise* goes to general quarters, it's sealed up from stem to stern. We had a lot of people from the air wing on the 03 level, just below the flight deck. Since we were on a twelve-hour basis, many of them would be locked in their compartments.

We had about fifteen or twenty aircraft on the after part of the flight deck, all loaded with fuel, ammunition, Zuni rockets, and bombs. We had this horrendous blaze going. There wasn't any doubt in my mind that the safety of the ship was paramount, and that we should go to general quarters, flood the hangar deck with the overhead sprinkler system, and try to prevent the fire from spreading from the after part of the flight deck to the rest of the ship. Once those fires got going with the minimum fire-fighting equipment that we had on the flight deck, there wasn't anything we could do about it. Also, metals like aluminum and magnesium begin to burn at very high temperatures, so piles of molten metal were growing back there.

Further, we had a number of Zuni rockets and 500-pound bombs loaded on various aircraft back in the pack at the after end of the flight deck. After a certain period of time in this intense heat, they cook off. When they cook off, they blow the airplane to which they're attached into a thousand pieces, sometimes off the ship and into the air, and sometimes blow a hole in the flight deck. The flight deck is a steel plate, perhaps two inches thick, of solid steel. However, with these 500-pound bombs going off two feet above the flight deck, they would blow holes in that metal.

We got turned into the wind, and I regulated the speed of the ship so that we would not have any smoke coming forward. We didn't need a lot of speed, so I just kept speed about 10 knots, and did the things that I've described earlier. There really wasn't any way we could fight the fire on the after flight deck. We just had to let it burn out, especially with those Zuni rockets, bombs, and ammunition. In the meantime, we were taking pictures of all this with the various TV cameras available to us in the bridge area so that we would have a complete record of what was going on. We were making every effort to prevent the spread of the fire, which was crucial. If the fire spread to the hangar deck, we could very easily lose the ship. We didn't want the fire to go below the 03 level. But there wasn't any way we could keep it out of there, and we had to man our damage control forces to those compartments that got blown through on the 03 level. We had to spray water on the bulkheads around them to make sure they didn't flash into flame that would spread on down below.

Burning fuel was pouring off all sides of the flight deck, so we had to keep water sprayed on those sides and on the fantail. The equipment that we did have all worked. Our people had just finished underway training, and we'd had hundreds of drills. That training paid off. We were able to confine the fire to the after part of the flight deck. After about three hours, the fire had burned itself out. We then headed in toward Pearl Harbor, and in the meantime, destroyers came alongside and tried to spray water on us and help us fight the fire. Their efforts were greatly appreciated, and took great courage on their part and great seamanship, but they just didn't have the water power to do us much good.

We had very few advantages fighting our fire as a result of what the *Forrestal* had gone through a couple years earlier. We had trucks on the flight deck with fire-fighting foam, a mixture of water and fire-retardant chemical, but it wouldn't have made any difference if we hadn't had it. We weren't any better off than the *Forrestal*, and the *Nimitz* (CVN-68) was no better off than the *Enterprise* during her fire in 1981. The changes that were made were largely cosmetic and ineffectual. What we really need is a dramatic change, a new piece of equipment. What you really need is a water cannon that can throw thousands of gallons of water per minute, or water mixed with a flame-retardant chemical, at the fire. Our sources of water for fire fighting are rather puny compared to what you really need. People manning fire hoses on the flight deck—that's ridiculous.

I was on the bridge the entire time, of course, monitoring what was happening on the flight deck, while our air officer was running the fire fighting going on, on the flight deck. I was also monitoring what was going on in the hangar deck to make sure we didn't have a fire down there. I was monitoring what was going on in the other parts of the ship. I directed that the water wash-down system be turned on up forward. We couldn't get to the valve to turn on the water wash-down system back aft, but it wouldn't have made any difference. That water wash-down system was like pissing into the wind, if you'll pardon the expression. We had settled the ship on a steady

Despite extensive damage from accidental ordnance explosions on the flight deck during operations in the Hawaiian Islands in January 1969, the Enterprise *was capable of limited flight operations within hours of the accident.*

course and speed, so there wasn't anything that needed my attention there. Also, I was in touch with the various destroyers around, to have them pick up people who had jumped over the side or to come alongside and spray us with water.

Of course, we reported this immediately to Pearl Harbor, and they sent helicopters out to take off our injured and dead. You're helpless up there on the bridge. There's just not much you can do to help anybody. I really didn't have enough to do on the bridge, because everything that could be done was being done by the commanders in charge of a given section. The engineering officer was running damage control, our air officer was running the flight deck, and our assistant air officers were running what was going on in the hangar deck. Our medical team, of course, was taking care of the injured. We had a very competent team handling all phases of it. I was primarily assuring myself that all the things that needed to be done were being done.

Some of the air wing personnel sealed in their spaces were among the fatalities and the injured because bombs cooked off above them; others got out. When we went to general quarters and they heard what was happening, they probably opened watertight doors and escaped, because if you look at the number of killed for such an inferno, it was surprisingly low. We had nine or ten 500-pound bombs go off high order. We had a lot of other bombs, but some of them didn't cook off high order. But we had only twenty-eight people killed.

Now when I say "only," I'm not being insensitive about it. We had a potential for killing hundreds, maybe even thousands, and we lost only twenty-eight. A higher death toll would have been a likely consequence had that space not been sealed off. We had to bend every effort to keep that fire isolated to the after part of the flight deck—keep it from getting into the living compartments and the hangar deck. A few men went over the side because the fire just covered the after part of the flight deck, and they were in a position that they had to go over the side. They were picked up by helicopters and destroyers.

After about three hours, the fire had burned out, and we began to count crewmen to see how many people were missing, killed, and injured. We had twenty-eight killed and 343 injured, with fifteen aircraft destroyed and seventeen damaged. We then headed for the Pearl Harbor Naval Shipyard. After any naval disaster, there's a board of inquiry. This is called in the vernacular "around the green table." Such came to pass after the *Enterprise* fire. The president of the board of inquiry was Rear Adm. Fred Bardshar. He had two or three lawyers to assist him. Other members of the board were specialists from around the Pearl Harbor area. The investigation went on for about five or six weeks while we were in the naval shipyard. I was around the green table a fair amount of time.

I was made an interested party to this board of inquiry, as were the air officer, the executive officer, and other senior officers responsible for the ship. As interested parties we were informed that we could have counsel, and we didn't have to testify unless we wanted to. We were afforded all our legal rights, in other words. I decided that we had done a good job in training our crew; I felt we had fought the fire very well; my only interest was to get to the bottom of the fire, find out just how it started, what went wrong. I decided to testify freely to the board of inquiry and not have a counsel; if I later needed a counsel for a court-martial, I would get one. At this stage the board of inquiry merely fixes blame; courts-martial would come later. I told my senior officers, who were also interested parties what I was going to do, but I invited them to make up their own minds about having a counsel. As it turned out, none of the officers on the *Enterprise* elected to have counsel.

The inquiry went off about as scheduled and, I believe, identified the cause of the explosion and fire. It was very simple. We had a nineteen-year-old driving a tractor with a small jet engine mounted on the rear. These jet engines produce a great volume of fairly high-pressure air that is channeled through a big hose and used to start a modern jet engine such as in an F-4 aircraft. These tractors are also used to tow jet aircraft around and consequently have a sizable diesel engine as motive power. This youngster had backed his tractor, with the little jet engine on the top of it, underneath the wing of an F-4 aircraft, on which were mounted Zuni rockets. As luck would have it, the exhaust of that small jet engine was coming out in such a way that it impinged directly on the warhead of a Zuni rocket. In a little more than a minute this hot exhaust cooked off the Zuni rocket.

The board of inquiry ran tests to see how long it would take one of these jet exhausts to cook off a Zuni rocket warhead. Each one took a little more than a minute. They cooked off regularly, never a misfire. It was decided that without a doubt this is what started the whole thing. The Zuni rocket warhead split into a thousand pieces when it exploded, and those little pieces of hot metal traveling at high speed were scattered all around the flight deck, wounding crewmen, puncturing fuel tanks, spilling JP-5 jet fuel all over the flight deck, and igniting it. In a matter of two or three seconds after the rocket exploded, the whole after flight deck was blazing with JP-5.

There were witnesses who saw the young man back the tractor up to the F-4. Several pilots had manned their aircraft; they were protected from the blast since they were in the cockpits. Fortunately, we did not lose any pilots. The explosion took place at a particular F-4, and the pilot was there. We had witnesses for all of this. We knew where the explosion started, and we knew that it was a Zuni warhead that went off. The pilot in this particular F-4 escaped, fortunately. The nineteen-year-old who was sitting in an open tractor was killed instantly by the shrapnel from the Zuni warhead.

The board of inquiry had great praise for the ship and the way the fire was handled. None of us was, to use legal terms, indicted. No one was court-martialed or got letters of caution or reprimand, or any other disciplinary action. The general feeling was that we had done about as well as we could do in a very difficult situation, that our crew was well trained, that we had fought the fire the way our equipment was designed to be used to fight fires, and that we had done all that was possible to do in such circumstances. Of course, that came as a great relief to all of us, because it had been a very traumatic experience. As we all know in the navy, the captain is responsible for whatever happens on his ship. I was fully prepared to take whatever disciplinary action was delivered to me. Much to my surprise, none of us were disciplined, and as a matter of fact, everyone—the crew, the officers of the ship—received great praise for the way we handled the whole thing.

I think the skill and courage of the crew were absolutely superb. In his report, Admiral Bardshar discusses how we fought the fire, and he commends the crew. All the equipment worked, the crew was well trained, and they showed considerable bravery. He especially had praise for the flight deck crews and the air officer who directed their efforts. I personally had no criticism for any aspect, or any person, or any phase of the fire fighting, and neither did the investigator.

After the fire was out we reached Pearl Harbor by the end of the day on the fourteenth. We went into the shipyard immediately, and the repair work started almost right away, and went on around the clock for six weeks. The shipyard did a great job. They weren't able to completely repair the ship, however. We had one elevator, one catapult, and, I believe, one arresting gear engine that were not repaired. But for all practical purposes, we had a fully operational ship.

We were very pleased with the final repairs and set to sea late on the afternoon of 5 March. Now this next tour to WestPac, second tour for me, only had one unusual occurrence. It wasn't very different from the previous three cruises the *Enterprise* had out there. We were going to the Tonkin Gulf and operations would be about the same. The only difference in this particular cruise was that an EC-121, an intelligence plane, was shot down by the North Koreans off the coast of North Korea in the Sea of Japan. Once again, the *Enterprise* was sent to the Sea of Japan to have a show of force opposite North Korea. It was winter and cold and miserable, and we stayed up there for a reasonable period of time, three or four weeks. It really wasn't very different from the *Pueblo* incident, in the sense that the Russians came down to greet us. We steamed back and forth and had flight operations in the Sea of Japan, once again, for two or three weeks, and then finally were relieved by another carrier and, once again, returned to Tonkin Gulf.

In 1969 we weren't sending the A-6s on night missions singly up to North Vietnam to bomb their capital. We were fully employed when we were in the Tonkin Gulf on various types of missions, bombing the trails of the North Vietnamese coming south, bombing trucks night and day. It really wasn't very different except that we weren't losing A-6s into the north the way we did on the first cruise. We had A-7s instead of A-4s for light attack on this cruise and it didn't make a lot of difference. The A-7 could carry a bigger load, greater accuracy, and better range. It was a good airplane; I think they were more effective, but there weren't many good targets out there. It was a very difficult war on all counts.

Our sources of intelligence were not all that good in North Vietnam. It was a guerrilla-type war, and our intelligence was very poor. The one area where our intelligence was better than most places was missile sites and the like; we did have good photography, and we could take our own pictures.

The locations of missile sites were pretty well pinpointed and kept up to date. But in terms of movements of troops, convoys, trucks, and the like, intelligence was very poor. It wasn't a war where intelligence was really used, not like World War II, where we used various types of intelligence to defeat the Japanese.

There was one item that I became greatly concerned about during these two years with the *Enterprise*—the maintenance problem of our aircraft aboard aircraft carriers. I made many hangar deck tours to see how maintenance was coming along, and it was just backbreaking. We just could not maintain the RA-5C, which was a dog of an airplane in the first place. The F-4s, A-6s, and A-7s filled the hangar deck on a regular basis with their need for repairs, the A-7s being a little better than the F-4s and the A-6s. But almost without exception, an A-6 would make one flight and have to go to the hangar deck for maintenance; the F-4 was almost as bad. Maintenance manhours per flight hour were running for those airplanes around forty to forty-five, which is just unbelievable. The RA-5C was a hopeless cause.

I decided that something had to be done in this area; we were failing. We had to carry altogether too many maintenance people on our aircraft carriers to keep these planes going. I decided that I would like to seek a job in naval aviation that had to do with procurement, maintenance, and reliability, and see if I could, over a period of my few remaining years in the navy, do something to improve the state of affairs. I applied for work in that particular area on leaving the *Enterprise*.

When you're flying combat missions twelve hours every day, there's a tremendous amount of paperwork that goes on, a tremendous amount of planning. You have to make decisions on weapons and tactics. Although every day we would get a message with our target assignments from commander, Task Force 77, we had to translate those into a flight plan and into our actions. Having a carrier division staff on board lessened the load on our ship. At first, I had some misgivings about having a staff aboard. But after operating out there without a staff a few times and operating, for the most part, with a staff, I don't think that the captain in the individual aircraft carrier is equipped to handle it all. What really happened during the Vietnam War is that the captain ran the formation and the operations of the ship in the task group. The admiral and his staff ran the war.

By and large, the staffs were very talented and very helpful. Without exception, the chiefs of staff were people who had command of a carrier. The operations officer was usually a captain who'd been a wing commander or something like that, so that the carrier division staffs were very helpful to the ship. I'd give them high marks; they made life easy for us. They handled all the messages; if we had not had a carrier division staff aboard, my officers, operations officers, air operations officers, would have had to communicate with Task Force 77 and Seventh Fleet, and make all the arrangements for all the strikes and all the planning. We were stretched thin. We really couldn't do it justice. With a carrier division staff aboard, they could handle all the communications with the Vietnamese and American types that we had to co-ordinate with; they could do all the debriefing and fill out the necessary action reports every day, and our people could help. It would then become a joint exercise. They would be there to debrief the pilots and make decisions or recommend actions to commander, Task Force 77.

I think one Vietnam tour, one cycle, is about enough during wartime. It's a twenty-four-hour-a-day job, and after one complete cycle, I was pretty well worn out. I was emotionally, physically, and mentally tired after the first tour. There was very little time to rest up; you got in the saddle and started all over again. After the second tour, I was doubly tired—mentally, physically, emotionally. It was very draining emotionally to see these people disappear, and to have to write the letters to their dependents. Two years of that was a really tough grind. At the end of those two years, and with the fire, I know I was completely exhausted. All I wanted to do was go find a corner and rest up for a month or so—get my perspective back. I think my effectiveness was impaired during the latter part of the command tour. I think I was a better commanding officer the first tour than the second tour, primarily because I was fresh and had lots of energy. The second tour, I think I was tired to start with.

I still believe that the best job in all the world is to be commanding officer of a ship; it's the best job there is. So under normal circumstances, I would say a two-year tour would be great, but with the war going on where you have twelve hours of flight operations every day, seven days a week, you get worn out, so that I think two years of that is too much. There are no breaks for the captain; even back in the States there were no breaks. A captain can take time off in those circumstances, but even though he gets away physically, the load is still there. You can't relax mentally, because the executive officer feels obliged to let you know of any untoward events.

The *Enterprise* got back into Alameda early in July, and I was relieved by my good friend, Capt. Forrest Petersen. I had been on the selectee list for flag rank for a year, so I made my number almost as soon as I was relieved on the *Enterprise*. I was ordered back to Washington and eventually reported to the Naval Air Systems Command as assistant commander for Logistics and Fleet Support.

19

Vietnam
The Final Chapter

Vice Admiral William P. Lawrence

William Porter Lawrence was born in Nashville, Tennessee, on 13 January 1930, son of Robert and Tennie Lawrence. He attended high school in Nashville, was appointed to the U.S. Naval Academy, and graduated and was commissioned ensign in 1951. He entered flight training at Pensacola, Florida, and was designated a naval aviator in November 1952. Subsequent service included extensive squadron, command, and staff duties ashore and afloat, and he advanced in rank to that of vice admiral, to date from 1 August 1980. Vice Admiral Lawrence was deputy chief of naval operations (manpower, personnel, and training)/chief of naval personnel from 28 September 1983 until his retirement in January 1986.

Admiral Lawrence's assignments included Fighter Squadron 193, the U.S. Naval Test Pilot School, and the Flight Test Division at the Naval Air Test Center, Patuxent River, Md. He subsequently served in various staff and squadron billets, and while commanding officer of Fighter Squadron 143, he was shot down over North Vietnam in June 1967 and held as a POW until March 1973.

After repatriation and convalescence, he attended the National War College, and after promotion to flag rank in July 1974, served as commander, Light Attack Wing, U.S. Pacific Fleet, director, Aviation Programs

Vice Adm. William P. Lawrence, deputy chief of naval operations (manpower, personnel, and training) and chief of naval personnel, 1983–86.

Division and assistant deputy chief of naval operations (air warfare), superintendent of the U.S. Naval Academy, commander, U.S. Third Fleet, and deputy chief of naval operations (manpower, personnel and training)/chief of naval personnel.

Admiral Lawrence's military awards include the Distinguished Service Medal with three Gold Stars, Silver Star Medal with two Gold Stars, Legion of Merit, Distinguished Flying Cross, Bronze Star Medal with Combat V, Air Medal with two Gold Stars, Joint Service Commendation Medal, Navy Commendation Medal with Combat V (two awards), and Purple Heart with Gold Star.

In June 1966 I joined Fighter Squadron 143 (VF-143) on board the USS *Ranger* (CVA-61) in the middle of a deployment to Southeast Asia and had the rest of the cruise out there flying the McDonnell Douglas F-4 Phantom. When I got there, we had stopped bombing in South Vietnam. We had built up the air force to the point that they could handle all of the South Vietnam bombing with their in-country assets, and so that's when the navy moved entirely to North Vietnam bombing. Before that there had been Dixie Station in the south Tonkin Gulf and carriers sort of warmed up there before they went to Yankee Station in the north. We did away with Dixie Station; it became all Yankee Station operations just before I got out there on my first deployment in 1966.

For planning purposes, North Vietnam was divided into six zones or packages—package one being the zone at the very southern part of North Vietnam, and package six the very northern part, comprised of the area of Haiphong over on the coast and then Hanoi inland. All of our targets were being assigned to us from the White House, so the target assignments would come out late the night before.

On the 1966 cruise, the majority of our targets were in the central part of North Vietnam, usually around package two or three; we really didn't get up there in the heartland of North Vietnam during that cruise. We returned to the States just before the big escalation occurred in the air war, and they started hitting targets that had been off the White House list. We really didn't see a great surface-to-air missile (SAM) threat in 1966. We knew they were there, but the Vietnamese had dispersed them pretty widely throughout the country.

The other thing that became painfully apparent to us was that we didn't have the electronic countermeasures (ECM) gear in our airplanes—the ability to break lock on the enemy SAM control system if a missile was charging up toward you. We also didn't have a SAM warning system in the F-4s. The A-4s had them, because the A-4 had the nuclear weapons mission where they would theoretically go over land in Russia, so they had a lot more sophisticated ECM capability than we had. We had nothing because the F-4 fighters, before the Vietnam War, were strictly used for fleet air defense. So we had to see the missile coming; fortunately, on that 1966 cruise, we really didn't see many missiles, because they were so widely spread through the country that they couldn't really mobilize very many in one place.

We had a flight schedule where one carrier, if there were two carriers out there, would conduct flight operations on the noon till midnight shift, and the other one would fly just the opposite. Sometimes we had three carriers and then one would fly midnight to noon; the other would fly noon to midnight; and then the third carrier would fly 8:00 A.M. to 4:00 or 6:00 P.M. The worst time to fly was from midnight to noon, because you'd wake up in the middle of the night and never get acclimated to normal sleeping. I hated the midnight to noon; I much preferred the noon to midnight.

The 1966 cruise, in terms of risk and everything else, as far as I was concerned, was very manageable. In other words, I knew I was being shot at, but I didn't really feel there was a great element of danger. We knew, of course, that the lack of electronic countermeasures was a real detriment to us. Fortunately, the skipper of our squadron, Doc Townsend, was very technically oriented, so he did a lot of good work in defining what we really needed in those airplanes to help us detect the missiles—a system that would detect the basic search radar signal and, then once the SAM did lock on you, could detect the guidance signal. So when we came back from that 1966 cruise, we had to put all of our airplanes through a modification program to get this ECM equipment. The system consisted of a strobe indication that would tell you the direction of the threat, and an audio signal to tell you when the missile guidance system had gone to high pulse repetition frequency (PRF) and was getting ready to lock on you.

Our squadron insignia that we painted on the airplane was a creature from mythology called the "winged griffin." It basically looks like a dog with a wing on its back, bent over with its mouth open. Somebody called it the "puking dog," so we were thereafter called the "puking dogs." But this nickname became a detriment; we'd fly all these missions, and at the end of the day the carrier public affairs officer would write up a news release to go back to the newspapers and wire services in the United States, but we had to send it down to the Seventh Fleet Detachment in Saigon for clearance. So every day we'd send out a message and our public affairs officer, or whoever worked with him, would state that the VF-143 "puking dogs" knocked out bridges and did all of this or that. Then the message would always come back from the Seventh Fleet Det, "To *Ranger*: your news release is approved except in line five, delete 'puking dogs, etc.'" No one would let us send "puking dogs" out in the news release.

The original Phantom II was going to be the AH-I, an attack airplane, and then the navy decided to use it as a fighter. So it had all the capabilities—wing strength, wing area, and wing stations—to be able to carry ordnance in addition to air-to-air missiles. So it was to be used very readily as a bomber. In the 1966 cruise we did a lot of two-plane bombing missions; sometimes we'd bring it up to four, but we weren't doing the large-scale, multiaircraft strikes, called Alpha strikes. Most of the missions I went on were either two- or four-plane missions in that 1966 time frame. Our squadron doctrine was that we would approach the coast at low altitude to within about fifty miles from the coast, where we would lower down to about fifty feet off the water, and then when we hit the coastline, we would pop up to higher altitude. Most of our targets at that time were within fifteen to twenty miles from the coast. After popping up we would then proceed in and make our bombing runs from 10,000 feet.

When we hit the coastline, we would have our speed up to about 450 knots indicated air speed, which gave us about 500 knots true air speed. So we'd be going pretty fast when over land to minimize the risk and then we would make our bomb run, usually a 45-degree run. It was a challenge, because we had no electronic navigational aids and had to do all our navigating to the target visually. You had to study that chart thoroughly, so that you could readily recognize the terrain features in helping you find that target, because it was really imperative that you know exactly where you were and be able to hit the precise roll-in point that you'd selected.

This was really a challenge at night. One of my real sore points about the Vietnam War is that we were sent in on night bombing missions with flares to try to find truck convoys, yet we weren't allowed to mine Haiphong or to cut the rail link to China—the northeast-northwest railroad. So we would let the supplies get in the country and then try to find the trucks at night, because they only rolled the trucks at night. It was the worst way in the world to prosecute a war. Those night missions were dangerous; the flares often didn't work, and when they did illuminate, they obliterated any horizon that you had and destroyed your night vision, making your instrument flying much more difficult. Then, of course, you couldn't really see anything on the ground.

In two combat cruises in Vietnam, I've yet to see my first truck on the ground, but I sure hurled a lot of bombs at the ground. We developed a tactic in our squadron working with the EA-3 Skywarrior electronic warfare airplanes at night. When the EA-3s would pick up some emission that would indicate to them that there was a significant target on the ground, they would inform an F-4 that was flying a mile in trail behind them. Then the F-4 would pop up, drop a flare, fly in a circle, come back, and try to get the target. But we were flying at an altitude of about 800 to 1,000 feet in trail on the A-3, keeping position on our radar—the ultimate in risk and demand of our pilots. Of course, our radar could be used for mapping, but it would not be able to discriminate a truck on the ground, so we used our radar purely as a navigational aid. It was very good to use that radar to pick out the place on the coast—what we called the coast in point. But that's the only way we really used the radar in the non air-to-air mode.

We actually got very good reconnaissance and intelligence support. The intelligence staff always kept an up-to-date chart on where the threats were—the antiaircraft and SAM installations—except that the Vietnamese kept moving the SAMs around, so that was not real valid. I tried to teach myself to use the Integrated Operations Intelligence Center (IOIC) on the ship, because there were some pretty significant capabilities in that intelligence center. For example, if you had a particular track that you were going to fly over the ground, you could tell an intelligence specialist, and he could get the reconnaissance photographs, put them in a machine, and you could actually see what your approach to the target would look like. It was a fantastic device. Only you usually were so hurried and time-constrained that you never had a chance to do that; but if you did have time, that was a great way to really get firmly fixed in your mind what would appear to you as you navigated to your target.

When you were leading a large number of airplanes on an Alpha strike, you had to be right on your exact track and bring the formation right to a precise roll-in point; everybody was basically flying formation on you, depending on you to get them to that roll-in point. It wasn't until you got them to that point that they started looking around where they were. So you had to be absolutely precise in your navigation.

We had some really close calls, because we were doing so much flying at night. Doctors did a study and found when they wired up pilots to take their vital signs during combat missions that when they came back to land on the carrier at night, their pulse rate was higher than when they were over the beach getting shot at, because night carrier operations are always very, very demanding.

One time when I was up on the carrier deck getting ready to go on a night mission over the beach, I was told there was a Russian ship up in the northern Tonkin Gulf that we needed to reconnoiter. I said, "God, just five minutes before launch, you're given this requirement, with no chance to plan for it." So my wingman and I launched and flew around at low altitude, trying our best to see if we could find this Russian ship or trawler or whatever it was. Of course, we couldn't find anything, but we were at low altitude, consuming our fuel excessively, and that bothered me, because at that time we didn't have in-flight refueling capability to the degree that we developed a little bit later on. So we came back to prepare for landing, and we were really getting low on fuel. As it turned out, we had to go in and land at Da Nang, because we didn't have enough fuel left. It was always touch and go going into Da Nang at night.

The maintenance and the flight deck people during that 1966 cruise were really great. I've always been so impressed how the American sailor can rise to the occasion. The pilots were on a twelve-hour flying period, but those maintenance guys worked as long as necessary. But, basically, the maintenance people were on twelve-hour shifts, and they worked those twelve-hour shifts for thirty days at a time. The weather was hot out there in the

Tonkin Gulf, and they weren't in the air-conditioned spaces, but they never complained, they always turned to. The flight deck crews, in particular, were doing heavy, physical-type labor. I was always amazed at how sailors would just lie down on a steel deck and go to sleep when they'd get a break. It's just wonderful, the type of young person we produce in this country that will really come through for you when you need them.

We've always had these real demanding jobs that we've given sailors. One of the toughest jobs in the navy, and it's one that usually goes to about a first class or second class petty officer, is what they call the final checker. He's the guy that stands up in the catapult area, and when one of his squadron airplanes comes up there, he looks all around the airplane for leaks or anything that would render it unsafe to fly. The wind and jet blasts are blowing over the deck and that guy looks around and comes out to give you a "thumbs up" to convey to you that the airplane's ready to go on that catapult. You always pick your very best guy, and I found over the years, invariably it's a midwest farm boy that gets that job. That's the type of guy that we produce out in middle America, and they're just unassuming guys that don't think anything at all about the danger. If an airplane turns abruptly and directs its jet blast wrong or if the blast deflector doesn't go up at the proper time, I've seen guys like that just blown right over the side and lost. But these guys that do that, that final checker, never manifest any fear and are totally dedicated and loyal.

Then, of course, there's the catapult crews. Every time you taxi up on that catapult, you know that your life is in the hands of those guys down there who are hooking you up and checking your airplane. You never have any doubt at all that they're going to be completely reliable; that's why there's such a great, good feeling of respect and affection between the pilots and the flight deck personnel, because we know that our life is in their hands. On the other hand, they have great respect for us, because we go out and do the dangerous stuff in the air. They're primarily aviation boatswains mates, really the salt of the earth just like the surface boatswains are too.

We had to just really knock ourselves out when we came back to the States in August 1966 to get all these ECM systems in our airplanes. This was done over at the Overhaul and Repair depot at North Island at Coronado, across the bay from San Diego, so it really reduced our capability to fly and train as much as we wanted to. The navy had a system by which the Bureau of Personnel was trying to get everybody in the navy combat experience, so after our aircrews had one Vietnam deployment, they were sent over to East Coast squadrons, and then we got people from East Coast squadrons coming back into our squadron. This was part of the concept of spreading the wealth, so to speak, and we got a big influx of people, either out of the training command or from the East Coast squadrons. We didn't have the airplanes available, so it was really quite a challenge to get the effective training done during that time between the end of that 1966 cruise and the start of the 1967 cruise.

I remember vividly how we hoped so much between cruises that the war would end. We were very much aware of this strategy of gradualism. It was very apparent to us, when we were out there in 1966, that the strategy was what some people called "squeeze them and wait," where President Johnson would bomb and assess the willingness of the Vietnamese to negotiate. The bombing pause in 1967 was at the time of their Tet, which was the Chinese New Year, a significant observance in Vietnam, like our Christmas season. So in Tet of 1967, Johnson ordered a bombing halt for a month. We at Miramar Naval Air Station really hoped that this would bring about the end to the war through negotiations. I remember how disappointed I was when we started up the bombing again, because we just didn't want to go back. We really had no eagerness to continue in the war.

It had been a very difficult time between August 1966 and when we went back in April 1967, with all airplanes getting modified. We had to go down to the Marine Corps air station at Yuma to do weapons training, then out on the carrier to do workups prior to the deployment. You really became aware of how different it is in wartime than peacetime. There's much more of a seriousness on the part of your aircrews and enlisted personnel. That's where I came to realize that when the war comes, a lot of the old traditions of aviator happy hours and things like that go by the board. I don't think we had a single happy hour in our squadron from the time we came back in August until we deployed again, because everybody said, "Gosh, boy, if I get some free time, I'm going to spend it with my family. I'm not going out and sit around the bar at Miramar." So it's a whole different attitude when you're in a war than it is in a peacetime situation.

When we came back from that 1966 cruise—I think it had been a nine-month cruise—we had two pilots that volunteered to go right back out on another carrier. I said, "I can't believe it; you guys have been out there nine months and you're going to take a month's leave and go back out on another deployment." They said, "Yeah, we really want to do it." I realized that some guys have this soldier of fortune mentality; there's something about the excitement of combat that appeals to certain guys, and these guys went back and did another nine-month deployment. They weren't married, and one of them had been my wingman, Terry Borne. He just loved every bit of it. After the next deployment they got out of the navy, and Terry became an American Airlines pilot. He truly had the warrior mentality. I could handle combat okay, but I tell you, I had no desire to go back out again. I didn't thirst for it, but there were guys that thrived on it.

We had a young pilot in the squadron, Ensign Brown. He was kind of a quiet guy, not really very outgoing, but we could see that he was a pretty good pilot and we were quite impressed with him. One week before we were due to deploy, he and his radar intercept officer (RIO) went over to pick up an airplane at the Overhaul and Repair facility at North Island. On the way back from North Island, he flew by the beach at La Jolla and at low altitude, high speed, did a roll and went into the water, killing both officers. It really just devastated all of us—as hard as we'd been working, getting ready to go back to war and then we were faced with this flathatting incident; we just couldn't believe it happened. We thought everybody was aware that you just didn't do things like that anymore. So on top of the burdens of getting ready to go back to war, we had to deal with that accident. It really demoralized our squadron.

We had to go through the memorial service; we didn't recover any of their bodies. Then, as we got under way from San Diego on the cruise, I remember the burden of having to prepare that Judge Advocate General (JAG) investigation and the accident report, which I was responsible for doing. It was really a very tough time emotionally; it took its toll on me to have to deal with that, and at the same time say goodbye to my family.

We went back to war on the *Constellation* (CVA-64) in April 1967. We had a real good squadron; there were four of us who had combat experience, and virtually everybody else in the squadron was new. But the very top leadership was combat experienced. Everybody was, from my perception, basically a very solid performer. I didn't have any trepidations about anybody's ability to perform effectively in combat. We had one period on the line in late May–June 1967, then we came back into Subic Bay in the Philippines and the commanding officer, Doc Townsend, and I had our change of command. Then we went back out, and I was later shot down midway through the second line period, 28 June 1967.

But things had gone well the first line period, and from everything I could see, our squadron was doing well. But it was a different level of intensity out there from the previous cruise. Basically, during our twelve-hour flying period, which would be from midnight to noon or from noon to midnight, we would usually fly two Alpha strikes, which would be about thirty to thirty-five airplanes each strike. There were no longer these two-plane or four-plane missions going after various targets in North Vietnam. We were doing all large-scale strikes. It was just another notch up in the escalation, in order to increase the intensity and the destructiveness of our raids. But still, we weren't striking the things that I really felt would hurt them.

For example, we weren't allowed to go in and wipe out the piers in Haiphong. We would strike a storage area back away from the pier, but we wouldn't go down and strike the dock area. We still were not allowed to mine Haiphong Harbor, and we didn't cut the northeast railroad leading from Haiphong up to China. So we raised the intensity, but we didn't raise it to the point where we'd really inflict significant damage on them. But it was very demanding for those of us who were executive officers and commanding officers, because we were planning and leading all these raids.

When you brought that many airplanes over the beach, you really had to plan very carefully what your route of flight would be, establish a roll-in point, and assign, within the target areas, specific targets for each group within the strike group. So the mission planning was very demanding, and it was made more difficult by the fact that the target assignments were coming out from Washington. Many times it would be midnight or one o'clock in the morning when we'd finally get the target assignments, after they'd come from Washington, and, apparently, out to CinCPac, CinCPacFlt, and then on down the line. Then you'd have to sit there with your rear seat guy and maybe one other, and plan this mission in the wee hours of the night, and the time on target might be seven in the morning. So, many times, when I was the strike leader, I'd finish the strike planning about three or four o'clock in the morning and get to bed for just a couple of hours. When you were the strike leader, once you planned the route, developed all the target assignments, weapons loads, and everything, you had to sit down and really study the map, because all the navigation to target was by reference to the ground. You simply had to be able to look down at the ground, recognize exactly where you were, relating the symbology you recalled from the map to what you actually saw on the ground. So I found I used to have to save a half-hour, where I'd sit down and study the map as carefully as I could so I could get a good mental picture of how the track to the target would appear to me from the cockpit.

The Vietnamese had the SAMs concentrated in the Hanoi-Haiphong area, and had steadily increased the amount of antiaircraft artillery in 1967. So on the raids that we were making into package six, the heartland of North Vietnam, the air opposition to us in all categories—air, missile, and guns— was very intense. Every mission that you took into what we called "Indian country" was really a tough challenge. Surface-to-air missiles put an entirely new aspect on the air war, because there was now an antiaircraft system where you could see the missile from time of launch. If you were a flight leader, as you approached the target, you had to keep your eye mainly on the ground, as I discussed previously. If you saw a missile lift off, you had to follow it carefully, because if it was locked onto you and your flight, then you had to maneuver very abruptly at the right moment to cause it to miss you. We had the capability to know when the guidance system was locked on us and also the direction of the radiation. But when you came into the Hanoi-Haiphong area, there were so many radars radiating that the system that gave you the direction of the radiation was of little value.

So you really had to keep scanning the ground, because you just simply had to see that missile from time of launch to be able to see if it was tracking you, so you could outmaneuver it. But the big problem was maintaining flight integrity with thirty to thirty-five airplanes while several missiles were coming up. We had several of our missions in which we just simply lost flight integrity. The airplanes became thoroughly dispersed before we got to target and aircrews had to make individual efforts to drop their bombs.

It was tough; we were kind of pioneering and developing our own tactics as we went along. We had what we called "Iron Hand" airplanes that had the capability to attack missile sites by firing weapons that homed in on their radiation. The Shrike antiradiation missile was the type of weapon used by the Iron Hand aircraft. But those missile suppression aircraft were not sufficient in number, and you knew if you went into the Hanoi-Haiphong area, you were going to get a lot of missiles thrown up at you.

One of the toughest demands of combat is programming adequate rest, because in addition to having to plan these strikes, participate in them, and go through the debrief, you had your duties during the day—running a squadron, being aware of the maintenance situation, taking your turn standing the alert watches up on the flight deck. We always had two fighters in what we called "condition one," to be prepared to launch in case of an enemy air attack. So you just seemed to have trouble squeezing out any time in a day to get caught up on your rest. I used to try to slip down to my room and get a twenty- or thirty-minute "battery recharge." But that was the real burden of combat for me—getting enough rest.

We had no problems maintaining our airplanes. Our maintenance personnel just did a magnificent job. We had very, very good availability; I never, ever remember having radar fail on a flight, ever having a communications problem; the F-4 was really a very durable workhorse.

Night carrier operations by then were a routine thing. Everybody carrier-qualified at night. But you had to be careful about your young pilots; you just didn't put them out in an unrestricted manner. You had to make sure the conditions were right—the weather wasn't marginal or you had an excessive pitching deck. You flew your young guys with a seasoned group of pilots until they started getting experience. The schedules officer tried to keep it pretty well even as to the number of night landings we were given, so that everybody would be able to maintain their proficiency. But I was quite impressed with what our pilots and radar intercept officers were doing on that cruise.

I had one RIO on whom the pressure of combat was starting to take its toll. I tried to have some long chats with him, where I was trying to stroke him and keep him going, and he made it through the cruise. I was really worried about him because he would come to see me and say that he was seriously considering turning in his wings. It was kind of ironic that after giving him as much encouragement and support as I could, then I got shot down. When I came back, he was one of the first guys to look me up. He said, "I thank you so much for giving me the strength to keep going and to hang in there during combat because if I had quit, I would never have felt the same about myself as I do now."

My last mission, the one in which I was shot down, consisted of over thirty airplanes, comprised of F-4s, A-4s, and A-6s. I was the flight leader and the navigational lead for the flight. Our mission in our eight F-4s was flak, or AAA, suppression—to drop cluster bomb unit (CBU) bombs, which dispersed fragments widely over the ground. That was the principal means that we used to suppress the flak, by impacting the people that were manning the guns and the missiles.

We carried air-to-air missiles, even when we were in an air-to-ground role, because after we completed the flak suppression mission, we would go up in the direction that we anticipated any air threat—MiG-21s or MiG-19s. We usually carried four Sparrow missiles and two Sidewinders, as I recall. In addition to those eight F-4s, I think we had sixteen A-4s, six A-6s, and the RA-5C photo aircraft with fighter escort. When we got airborne, everything we did after launch until we were finally joined up in a big circle over the carrier was on radio silence. We really prided ourselves that we could launch all those airplanes and rendezvous in their individual four-plane divisions without ever making a transmission on the radio. Then once we headed into the beach, there'd be maybe one transmission that indicates everybody was on board.

Our target on that particular mission was the transshipment points in the port city of Haiphong. After the supplies were offloaded from the ships, they were moved back to a transshipment storage area where they were picked up by trucks. We were allowed to attack those storage areas, but not the piers. If there were foreign ships at the pier, we had orders to be very careful about damaging them. Apparently, President Johnson was very worried about antagonizing the Russians and the Chinese and maybe precipitating their entry into the war.

As we were heading into the target, before we got over land, I could see that there were very heavy cumulus cloud buildups over Haiphong, even though it was early in the morning. As we got closer, it was apparent to me that we could not make satisfactory bomb runs. Our alternative target down to the southwest was Nam Dinh. When I saw this cloud cover over Haiphong, I turned and took the flight toward Nam Dinh. Our targets were the thermal power plant in Nam Dinh and the transshipment areas adjacent to the Red River, which led down from Hanoi out to the ocean.

Now at that time in the war, because of the repeated missions into the Hanoi-Haiphong area, the enemy had brought missiles in from all over North Vietnam and ringed them tightly around Hanoi-Haiphong. I remember when I turned down to go to Nam Dinh, I thought, "Boy, I won't have to sweat the missiles today, because we'll be outside the missile envelopes." So what happened? I got shot down by AAA—a World War II 85-millimeter gun. I was going 500 knots at 10,000 feet, so that was a darn good shot for a guy to sight on you at ground level, and fire a shell up to hit you at 10,000 feet going that fast. Some of their guns had fire control radar and some didn't. I don't remember getting an indication of being radiated before I was hit.

I was really concentrating very much on keeping myself precisely located by reference to the landmarks on the ground and I was constantly turning the airplane, which we called "jinking," to make it difficult for the enemy to sight on you. So you were the literal "one-armed paper hanger" as a flight leader, because you had to make all this stuff come together—you had to bring all these airplanes into a precise roll-in point, enabling them all to make a satisfactory bomb run. You had to have a real good sense of judgment, because we were coming in at very high speed, and couldn't afford to circle around the target to reach your roll-in point, excessively exposing the flight to much greater enemy fire. So we had to perform what we called a "roll forward"; at the proper moment we would pull the airplane over on its back, and then pull down tightly into our bomb run.

As I was within about five to ten seconds from my roll-in point, my airplane was enveloped by flak, marked by many small puffs of smoke, and I felt a pronounced jolt in my airframe; but the airplane seemed to want to keep flying. My wingman, Lt. Tom Rodger, called up and said, "Hey, skipper, I think you've been hit." I felt like saying, "Tell me something I don't know." As I got closer to the roll-in point, I saw the hydraulic pressure warning lights begin to flicker, but the controls still felt pretty good. So I made a snap decision not to pull myself out of the flight and head back to the ocean, but to complete my bomb run.

As I rolled onto my back, and settled into my bomb run, I could see out of the corner of my eye that the warning lights for the two flight control systems and the utility systems all were steady red. Apparently, lines in all three hydraulic systems had been severed, and as I was pulling my nose down, I could feel that the airplane controls were starting to get kind of mushy. But, with difficulty, I was able to get my bomb sight on the target and release the CBUs. I didn't have to be all that precise, because CBUs are an area weapon; fragments are spread all over the area, unlike conventional iron bombs.

After bomb release, I had a hell of a time getting the nose of the airplane to come back up. The controls were unresponsive, and I almost flew into the ground, because the airplane was recovering from the dive very sluggishly. As I started to climb, I immediately headed back toward the water. I got to about 10,000 feet, and the airplane just went completely out of control. I guess, finally, the hydraulic pressure just went to zero. The aircraft settled into a steady, flat spin. My controls were having no effect, so then I went into full afterburner on my right engine and back to idle on my left to stop this right rotation; but that didn't have any effect.

At about 3,000 feet, it was obvious to me that it probably was not going to recover from the spin. So I told my rear seat guy to eject. Around 1,800 feet, I knew I was in a perilous situation, so I then ejected myself. We always carried hand-held survival radios, a kind of "walkie-talkie" radio, so as I was coming down in my parachute, I was able to get that radio out of my integrated harness and transmit to the other aircraft that I was out of the airplane and was okay. That was fortunate because that gave U.S. authorities the feeling that I probably had survived and was a POW, although my name was not released by the North Vietnamese until about three years later. My RIO was Lt. (jg) James W. Bailey, USNR, who was twenty-four years old at that time, and luckily unmarried. I saw his chute about half a mile away from me as I was coming down, so I felt good about that, and I reported that on my radio. So our squadron mates were very confident that we both had survived our shootdown.

I think I had a good chance at being rescued if, when I was initially enveloped with that flak, I had immediately turned back toward the Gulf, since we were only about twenty miles inland. But I never really seriously considered it, because the airplane felt like it could complete the bomb run. All your background and training tells you that you should complete the mission, and, of course, as a flight leader, I couldn't very readily pull myself out without disrupting the flight. I never dwell on that very much, but I have to admit that during my first several weeks as a POW, I went through this "Why me?" period, where I thought I must have done something wrong to facilitate their being able to hit me. For example, I used to think, "Gosh, was I jinking enough?" I couldn't remember whether I was really jinking as much as I should. Later on, when I got released, Roger and the other guys said, "Yeah, you were jinking and doing everything right." It was really a hell of a good shot for the enemy gunners. On the other hand, it may be that I ran into the shell rather than the other way around. Of course, being the lead airplane on a clear day, you're very vulnerable, and that's why so many commanding officers and executive officers got shot down in this war, I think, because we were always the flight leaders.

I'd been shot down over the Red River Delta, which is a major rice-growing region in North Vietnam. I landed right in the middle of a rice paddy, and the water was up to my thighs as I was trying to get my bearings and prevent my parachute from entangling my body. There was a militia man standing over at the edge of the rice paddy pointing an old rifle at me, so I had no chance to try to evade and escape, since I was captured right at the instant I landed. I was taken immediately into custody, and it was obvious that this guy was very inexperienced at this; I was probably the first American he had ever seen face to face. In the delta, there are thatched houses called "hooches" interspersed throughout the rice paddies, and they form little hamlets. My captor put me into a hooch, which was a pigpen containing a big sow pig. She must have weighed 400 or 500 pounds—the biggest pig I've ever seen. The guard had me stand over in the corner, and I noticed the whole time I was in there, she kind of was eyeing me, like, "What in the hell are you doing in my bedroom?"

Later on he came back with somebody that seemed to have more authority and this person took me in custody. One of the first things they did was to have me strip down. I took off my integrated harness, pistol, hand-held radio, boots, and my flight suit—essentially stripping me down to my skivvies. I was barefoot, and they looked in my mouth to see if there was any gold fillings. Of course, they took my watch; I always left my Naval Academy ring back in my stateroom. They really stripped me down to just the bare essentials.

Then they started running me across the countryside to the place where we were ultimately picked up by a truck. Jim Bailey, who had landed about half a mile away from me, was taken separately to this same collection point. It was kind of interesting, as we ran through these little hamlet areas, to observe the attitude of the Vietnamese people. The young kids thought it was a big social event. They laughed and cut up, and the only people that seemed to have any real hatred were the older people, probably those that had gone through the first war against the French and had experienced French colonialism. They were the only ones that stepped forward and tried to strike me. Most of the people there that were in their twenties and thirties were essentially passive. The guards had me trot until we finally reached a collection point. I'd been shot down right around 7:30 in the morning, which had been our assigned time on target, so it was still quite early.

At the collection point, they blindfolded us, tied our arms behind us, and put Jim and me in the back of the truck. A guard rode with us, and would not allow us to talk to each other. If we tried to whisper or anything like that, he'd hit us with the butt of a rifle. It took us a fairly long time to get up to Hanoi. I guess Nam Dinh, the point where we were shot down, is around twenty miles south of Hanoi, but it seemed to take us an extraordinarily long time, because the roads were full of potholes, requiring the truck to travel at low speed.

It was well into the afternoon when we finally arrived at Hanoi, because we stopped at one intermediate point and were taken into a militia station. A uniformed official put before me a questionnaire, in English. The questionnaire required a lot of information other than just name, rank, and serial number. So I refused to give any additional information as required by the Code of Conduct. Surprisingly, the guy accepted it; he didn't really press me.

So I said, "Gee, maybe this is the way I'm going to be treated. Maybe they're going to abide by the Geneva Convention." But then when we got up to Hanoi, I found it was a different ball game. When I finally faced the North Vietnamese officials, it seemed to be in the early evening; it was getting dark. They had brought us into a large prison, which subsequently I learned was called Hoa Lo and has become known as the Hanoi Hilton. The area where they took the new prisoners for the initial interrogation was what we later nicknamed Heartbreak Hotel. Since there were so many air force officers, a lot of the places in that prison are named after casinos in Las Vegas, because many of them had trained at nearby Nellis Air Force Base. We had a name for each prisoner compound and the interrogation rooms—Gold Nugget, Thunderbird, and Stardust were some of the names of the locations in the camp. Our entire camp was called Vegas.

I lived initially in the interrogation room, which contained only a single table and several chairs. The room was totally isolated from the main camp. They initially brought me before three officers, one of whom spoke English. He started asking me various questions and I replied that I was only required to give name, rank, serial number, date of birth. He kept insisting that I was an air pirate and a criminal, not a prisoner of war, so I would not be accorded the rights of the Geneva Convention. He kept asking what was the ship that I was flying from, what was my target, type of airplane—those types of questions. He also wanted detailed information about my mission. Of course, another thing they were interested in was what would be the future targets that would be hit. They were seeking basically military tactical information, as opposed to other types of intelligence.

Well, I kept refusing, and then, finally, after several confrontations, discussions, and so forth, the officers left the room. Then a guard that we came to know over the years as the professional torturer came in; the name we had for him was "Strap and Bar." We nicknamed every guard that we had in the camp, and, surprisingly, we had essentially the same guards the whole time we were there. Apparently, that was their assigned duty in the army, and they just stayed in those roles. This professional torturer, "Strap and Bar," was really kind of unreal. He was very laid back and phlegmatic, but very efficient at what he was supposed to do. We POWs speculated why they picked a guy like this to be the torturer, as opposed to some vicious type of guy. I think that they realized that they had to have a very skillful guy because an overzealous torturer could easily seriously maim or kill prisoners. They knew we were valuable as hostages and they didn't want to dispose of us, although we know that there were prisoners that were probably killed through torture by brutal, inept guards.

This guy was very professional. He had straps and a bar, and he would, essentially, put your body into contorted positions, such as placing your legs into shackles on a long, horizontal bar, and then pushing your head over underneath the bar, so your body was very much bent and compressed, inducing a high level of pain. Then he tied your arms behind you, and left you in this very painful position. I think all of us considered ourselves James Bond types who could resist to death if we had to in order not to give information. But we all learned that one can take pain for just so long, and eventually has to think of another strategy. They didn't give us the option of dying.

I tried to endure as long as I could; I probably passed out several times, because I remember I also had a type of a strap around my neck that was cutting off the blood flow to my brain, making me black out. This continued for several days; it's kind of dim in my memory, because I was in and out of it.

The type of information they were seeking from me was what were the planned targets, which I didn't really know, because the targets were assigned from Washington. Finally, I realized I had no other alternative, that I had to tell them something. So I just started giving them targets that would cause them the most damage, like the piers at Haiphong, the government palace in Hanoi, and their jet airfields. Of course, I had no idea of the validity of what I was saying, but it seemed to satisfy them and got them off my back.

I was really worried that they would ask me about some very sensitive intelligence information, because I had some very highly classified knowledge about some electronic countermeasures systems. Surprisingly, the questions were not sophisticated at all; they were just very simple questions. Our torturer was so competent and effective that they probably could have gotten more from us.

I think their objective was to show that they could make us talk, rather than trying to get specific information, but, also, I think it was that they were not very sophisticated. And that's where I came to realize that the Chinese and the Soviets, or any other foreign power, were not closely involved with the North Vietnamese. These powers would have been urging them to get more advanced information from us. Another point was that brutal torture seemed out of character for the Vietnamese people. They're not a fierce culture, like the Koreans, for example. We only had one or two guards that you could consider brutal. And the others—I wouldn't say they were kindly toward us, but they never really were extremely vicious in their actions against us. But if they were instructed to punish you, they would do it well, because that was their job. I guess they might have been motivated by a fear that they would be disciplined if they didn't punish us.

It was obvious that there was an attempt by the leadership of the North Vietnamese to keep their people incited all the time—rising up to fight the enemy, a foreign power that's trying to recolonize them. Even in spite of that, the guards rarely got real vicious with us. But, as I say, if the camp leaders were determined to make you do something, like make an antiwar propaganda statement, things like that, they would efficiently apply the measures to do that.

I was fortunate, I think, because I held out pretty hard and established my credibility as a tough resistor, so they never tried to exploit me for propaganda, such as meeting with a Jane Fonda type of visiting delegation. I was punished a lot, and put into leg irons and isolated in solitary confinement, but it was mainly because they knew I was running camp communications and serving as a leader. They never sought sensitive information from me, and I was quite perplexed that they wouldn't be interested in more military data, particularly our bombing tactics and our procedures to counter their airplanes. The only other time they tortured me for information was when they wanted all POWs to fill out a very comprehensive questionnaire on personal information, like members of our family and life experiences. There was nothing in there that I saw would be harmful to the United States, but as a matter of principle, I resisted, to make them have to work to get anything out of me and keep my credibility up. There were many cases of POWs being exploited for propaganda purposes, but I was fortunate that I never was put in that position.

Our view of visiting U.S. delegations was very negative. The most prominent were Jane Fonda and Ramsey Clark, who received the most publicity. Jane Fonda made a tape that they played to us basically saying, "The U.S. government's wrong" and "You've been misled" and "The war is unjust." And we heard many antiwar statements from Ramsey Clark. I feel Fonda and Clark were just hopelessly naive, and that they had no concept of the nature of the enemy and what communism was all about. The thing that I resent mostly is that Jane Fonda, although she has admitted that she perhaps made a mistake in visiting North Vietnam, has never really taken positive action to overcome what she did. Ramsey Clark was the one that came across to me as just being hopelessly naive, and I was amazed that a man with his background would have gotten so far off the track.

After that initial torture session, I was really banged up and I couldn't use my limbs very effectively. It took several months for me to heal from that experience. I had a fractured bone in my left shoulder and I'd totally lost the feeling on the outside of my lower right leg due to nerve damage. I was taken into the main camp, which we called Las Vegas, or Camp Vegas, and I initially went into the compound we called the Thunderbird. I was put into a cell by myself, but very soon I could hear POWs talking. I had to pull myself over to the door, because I was still somewhat crippled from the torture. I did get involved in some talking, and I found out that our senior officer in that compound was Jim Stockdale. He and I were old friends from our test pilot days. I was able to pass to him that I had seen his wife, Sybil, about four months before I was shot down, at the change of command of the aircraft carrier USS *Constellation* over at North Island in San Diego. I was also able to tell him that his name had just appeared on the captain's promotion list, so that was a boost for his morale, because he hadn't heard anything from his family.

In the course of this talking, they did pass me the tap code, and then I moved over to a room where I had a common wall with a POW named Ron Maston, who was an air force lieutenant, so he and I started tapping to each other. It was amazing to me how quickly you can become proficient in that tap code; even in those early days I was able to transmit a fair amount of information to Maston.

But it was tough getting acclimated to the prison life—the austere conditions, sleeping on a bare, wooden bunk, the heat, and the lack of sanitation. I soon started getting very bad sores in my scalp. At that time, instead of rice, they were feeding us coarse, brown bread, because I think they imported flour from other communist countries. I remember one night I had placed my uneaten bread on the bunk, and later a big rat began pulling this piece of bread. The rats there were just unbelievably big—some of them were as much as a foot long. In later times when I was down trying to communicate under my door or through a water drain in my cell, I'd look up and be face to face with a rat. The hygiene was incredibly bad, and I think that was just the way people were accustomed to living over there. It wasn't just within our prison. I think that was a fairly prevalent standard of hygiene throughout the country, and it was tough acclimating to those conditions.

We had very limited medical care. One thing the French had done during colonial times was to establish a rudimentary medical system. I was fortunate that I never got seriously sick as a POW. The worse thing that happened to me was a very bad eye infection; they took me out to our camp corpsman in his small, filthy clinic. He used a big needle to gave me an injection, which, in retrospect, I think may have been penicillin, because I broke out in a very bad rash, since I am allergic to penicillin. Luckily, that was the only time that I required medical care. When I was shot down, I was in tip-top condition, because I'd kept myself physically fit. I think that really helped me get through the experience, because there were a lot of guys that were sick the whole time they were POWs with dysentery or some malaise.

They kept me in solitary confinement from the time I was shot down, which was the twenty-eighth of June, through that summer, and they moved me from cell to cell from time to time. After I'd been there about a month, the POW communications abruptly ended in Thunderbird. I later learned the camp commander had discovered that Stockdale was putting orders out to the other POWs. Somehow they obtained this feedback through POW interrogations or other means, so they took the twelve POWs that they perceived were the main leaders of the resistance, and isolated them in a special camp called Alcatraz. When they pulled all those guys out, it just ended communication in Thunderbird, so I didn't communicate with anybody from about July to October.

I was in solitary confinement, but I was trying to do everything I could to see as much as possible. One time I was looking out my window in this one cell and I saw my Naval Academy classmate Byron Fuller, being helped by another POW. It was the common practice that if we had one POW who was badly injured or sick, another POW would be assigned to be his nurse. So Byron was being helped out by his nurse, and I said, "My God, that's the third guy out of my Academy class that's been shot down," because Al Brady had been shot down about a month before me.

Later, when I was looking out, I was detected by a guard. That's the first time I got put into leg stocks, which were basically shackles at the end of the bunk. This happened to me after I was there about two months, and it was real demoralizing. I was in a small cell and it was terribly hot. That was the only time in the POW experience when I really got low, because it was now obvious they were not going to let me write or receive mail from my family. Because of the discomfort of being in leg stocks and the heat and everything—I really became demoralized, but somehow I was able to pull myself out of it. That was the only time in my whole six years I ever had anything like depression, because most of the time I was actively trying to resist, to keep in contact with the other POWs, and those challenges gave me something to keep my mind off our bad conditions.

That period in the stocks lasted about a month, and then in October 1967 President Johnson significantly escalated the bombing and we had massive air raids in the heartland of North Vietnam. I remember several times a day we could hear the air raids coming in, always preceded by the air raid siren. As a result of these intense raids in October, there were quite a few American airmen shot down, so when they decided my punishment was up, I was put together with three other new POWs. One was my Academy classmate Chuck Gillespie, another was Tom Kirk, and the other was Verlyn Daniels. They had all been shot down on raids into the Hanoi-Haiphong area. Chuck Gillespie was the commanding officer of Fighter Squadron 151, Tom Kirk was an air force lieutenant colonel, the operations officer of an air force fighter squadron, flying F-105s, and Dan Daniels was the executive officer of an A-4 squadron.

It was a great improvement having roommates after a long isolation, and we basically stayed together on and off until the end of the war. I went through long periods where I was just with Tom Kirk, and then I experienced some more solitary confinement. My total number of months in solitary confinement over the six years was fourteen months. After joining with those guys, I learned I was the senior in that group, and I said, "We've got to get a communication system going in our compound." I told them the tap code, and then we started doing a limited amount of talking with other POWs in our compound, because we figured out a way to look for the guards. We knew that communications were absolutely imperative to have a military organization and an effective resistance, but also to help POWs keep their morale up.

So, after really analyzing the situation, we realized that the guards, during certain times of the day, had kind of stereotyped patrolling patterns as they moved from one compound to another in Camp Vegas. There were three large and two smaller compounds in Las Vegas. By looking out of a crack around our door and window, we could tell when the guard departed our compound, and we knew that it would be about ten minutes before he came back. That was the time we talked to POWs in adjacent cells, mostly during the siesta period, from about 1:00 to 2:00 P.M. I guess the French had introduced them to siesta, and the North Vietnamese kept it as a part of their culture. That's the time when the camp got pretty quiet, and they expected us to nap; instead, we did most of our verbal communications then.

We finally got a pretty good communication system established within our compound, both by talking and by tapping, and that's when I took effective command of the camp as senior ranking officer. I kept that command position for about the next two years, because Stockdale, Denton, and other senior POWs were all in Alcatraz and other camps. We also found that we could take stiff bamboo brooms they gave us for cleaning, and by using strokes of the broom, we could send the tap code.

The tap code we used was based on a matrix that we formed by taking the twenty-six letters in the alphabet, dropping off the letter K, and making C and K interchangeable, leaving twenty-five letters, which were put into five lines of five letters each. The first line was A through E, the third was L through P, and the last was V through Z. To tap a letter in that matrix, you would tap down the left side, then to the right. For example, the letter M was the second letter in the third line, so you'd just go "tap, tap, tap," pause and then "tap, tap."

In addition to sending code by tapping on the walls, you could also do it with strokes of the broom. By pressing your ear to the wall, you could hear light tapping as much as twenty to twenty-five feet away and we were able to relay messages from cell to cell. So over time we built up this fairly sophisticated communication system, based on limited talking and extensive use of the tap code. We also used typical sounds you make when you have a respiratory ailment to transmit the code. For example, the first line would be one cough; the second line would be two coughs; the third line would be a throat clear; the fourth line would be a very deep hack; and the last would be spitting. We would send brief messages with the voice tap code if we had POWs that were in locations where we couldn't tap to them on the wall. For example, a message we'd usually send would be something like GBU, which stood for "God Bless You."

To keep the guards confused, we would be constantly sneezing and coughing, just to mask the times when we wanted to send that voice tap code. Also, we were always bugging them to let us sweep out our cells, because that would be a chance to send messages, by sweeping with a broom that could be heard all around the camp. So they thought we were absolutely nuts on the subject of cleanliness.

We were in a constant confrontation with our captors, because they were always trying to find some prisoners that they could force to do something, like make a propaganda statement. There were two POWs who actively collaborated with the North Vietnamese on making propaganda statements. One was in my compound and once when I was talking with him, he suddenly just went off the line, and I just couldn't get him to communicate after that. Apparently he had made his decision that the pressures were so great that he was going to collaborate with the enemy to have a better life. I remember trying my best to keep him communicating, to no avail.

It was a real challenge to keep everybody on the line communicating. One of the first signs when a guy was having some problems with depression would be that he would stop communicating. You just couldn't get him to come up, either to tap or to talk. Another sign was when a guy stopped eating, and we had to use a lot of means to get these signals. We were in individual cells—about seven feet square—and as you finished up your meal, they would have you bring your little bowls out to put in one place. So if I came out and I saw a bowl that was full of food near a door or something, I would say, "Better start watching that guy; he might be having problems." If a guy just lost his appetite, it was often a psychological problem.

The guys went through periods where they would be very, very depressed and just didn't want to communicate. And, also, there was this great fear of being caught communicating, because the guards would really work you over in those early years prior to about 1970. If you were caught communicating, boy, you really got punished, so a lot of guys didn't relish the thought of communicating and getting caught. I had to really be pretty firm as a senior officer requiring people to communicate, because I knew that if we didn't communicate, we couldn't have an effective resistance. Also, I knew guys would be psychologically healthier if we kept them communicating. If you didn't work at it all the time, you just couldn't get the information passed around; it was so hard, with the guards patrolling, to find those times to communicate. So you had to exploit every opportunity.

The Vietnamese tried hard to discourage communication because they knew that this would enhance our resistance, and they really wanted to exploit us to the absolute maximum for propaganda purposes. I think they had this feeling that if they could get prisoners to say we were against the war, this would be a valuable propaganda tool. To my knowledge no one under my command while I was a senior officer ever went out to see a visiting antiwar delegation. We had a pretty tough resistance posture, and our captors must have regarded Camp Vegas as not being a potential for exploitation.

As best I could determine, only one POW died in the camp while I was a

senior officer—Lance Sijon, who was shot down as an air force first lieutenant. After he'd been there a while, I understand the air force promoted
him to captain. He had evaded for a long time after being shot down and he
was suffering from very advanced exposure, with a broken leg and severe
gangrene. He was brought into our camp, and eventually died there. Even
though I was never in direct communication with him, I learned through the
communication system that he was in very bad condition. The guards eventually took him away and he did not return. After the war, he was posthumously awarded the Medal of Honor, and there is a hall named for him out
at the Air Force Academy, which he attended. There were also other prisoners who were taken out, apparently in good health, and just never returned.
After the war, we learned they had died in captivity. We never really learned
all the circumstances in these cases.

Over the years we realized it was necessary to promulgate amplifying
guidance for POWs to that contained in the Code of Conduct, specifically
with regard to torture. We learned that we could not endure constant severe
pain until death and that it was best to have a preplanned cover story or fallback position. Before losing your rationality or being permanently damaged
mentally or physically, it was necessary to skillfully deceive your captors
without giving them anything of great value. We issued instructions that
POWs should not take the fallback position until they had endured significant pain. By hanging in there as best they could, they would establish credibility with their captors as a "tough nut" and the Vietnamese would be less
prone to come back again for exploitation. Another reason for taking significant pain was to salvage your self-respect, which was perhaps more important than anything else. After giving in to sustained torture, a POW naturally
felt very low, but it was easier to "bounce back" if you had given it your best
shot. Even though you had some residual injuries and pain, you had a much
better feeling about yourself because you hung in there rather than capitu-
late right away.

I found that individuals' tolerance of pain varied significantly. Some
POWs absolutely dreaded the thought of being physically roughed up. So
you had to spend a lot of your time, as the senior officer, trying to boost people's spirits and reinforce their ability to endure the ordeals of captivity. I
spent more time as a cheerleader than giving specific orders.

When the Vietnamese released some POWs in the summer of 1969 as a
propaganda measure, these guys went back and revealed all the torture and
mistreatment that was going on. That release, which the Vietnamese thought
would get them world sympathy, really backfired on them. They were so
criticized in the international press that they finally started in late 1969 back-

ing off on the physically harsh torture. But they still punished you, particularly if you were caught communicating.

One type of punishment they would give was just to have you sit in a chair. Initially you would say, "Golly, that's not very difficult." But then after about twelve hours or so, as you're sitting in there, the blood pools in your ankles, you don't get any sleep, and you eventually start hallucinating. Just sitting in a chair for days really got to be very painful.

After the treatment improved and we became more clever and skillful in communicating, we began passing notes around the camp. We established a system of note drops or "mailboxes" in common places where POWs would go, like the outside bath stalls. I got caught one time putting a note in a note drop. The Vietnamese knew that I'd been one of the leaders of the resistance all those years and I think they finally just decided they were going to break me. So they placed me in a six-foot-square cell that was unventilated, unlighted, and had a tin roof; we called it Calcutta, from the Black Hole of Calcutta. There was hardly any room beside my wooden bunk to permit walking. It was August, and during the day the temperature would go up well above 100 degrees. In the summer months, those of us with fair complexions would have heat rash over our entire body, but on that occasion, the heat rash advanced into painful heat sores over my entire body. I had to lie completely immobile on my back to minimize the discomfort.

I soon realized that I had to get a deep thought process going to keep my wits. Of course, I'd known from my early years as a POW that when you're in solitary confinement, you couldn't daydream and fantasize; you had to have a good productive mental process or you would become literally a vegetable and your morale would deteriorate. In my early captivity, one thing I did was to relive my life in great detail, and I did that about three times. It used to take me almost three weeks to go through this process in my mind, trying to resurrect as much as I could remember, starting with my earliest memories.

My earliest memory was as an infant in a crib adjacent to my parents' bed. One thing I did was to try to recall as many persons as possible from my school classes. I would concentrate on such an effort for hours on end and then the names would start appearing in my mind. It was amazing to learn the working of the mind.

When I went home to Nashville for the first time after being released, people would come up to me and say, "Well, I'm so-and-so; I was in your elementary school class." I would respond, "Yeah, I remember you," because I'd resurrected their name in captivity. What you basically did as a POW was to shrink your world down to where, at times, it just encompassed your cell; you no longer thought of external affairs, and your objective was to get through that particular day in the most effective way possible, involved in productive thought processes.

I'd review history, do complex math in my head, as an example. During one period I even designed houses in my mind. When feasible, I'd require myself to do calisthenics, even though it was hot and I was physically debilitated. Sometimes I would pace back and forth in my cell for long hours in solitary confinement. But on this one occasion in Calcutta, I knew that I had to use extraordinary measures to get my mind off my intense pain. Fortunately, the guards were leaving me completely alone, since the Vietnamese had ceased the harsh, abusive treatment.

I finally conceived the idea of composing poetry in my head. I had remembered from my tenth-grade English, a great poet, Sir Walter Scott, who, using the iambic pentameter poetic meter, composed *Lady of the Lake*. A typical line from this poem was "The stag at eve had drunk his fill, where danced the moon on Monan's rill." I made as my goal to compose a perfect iambic pentameter poem. Sir Walter Scott had genius, but I had time, and I was determined to do it. So I started really thinking deeply, and I came up with the line, "Oh Tennessee, My Tennessee," as a perfect iambic pentameter line. With that as a first line, I went on to compose a poem that expressed my feelings about my native state. I set as my goal to have every line in there describe something special about my native state. It took me about three weeks to compose that poem in my head, deeply involved in thought for about sixteen to eighteen hours a day. It really helped me get through a difficult period. Then I composed some other poems, and I kept them stored in my head after that.

I finally got out of Calcutta after two months and was brought back into the camp and put back in with my old roommate, Tom Kirk. I found out later on that the POW who had succeeded me in Calcutta was John McCain, after he'd gotten into some trouble.

When I finally was released from captivity in 1973, one of the first things I did when I was in the hospital at Clark Air Force Base in the Philippines, was to sit down and write out all these poems. When I arrived at the naval hospital at Memphis several days later, I was informed that my hometown of Nashville planned to have a "welcome home day" for me. As part of that program I was expected to make several speeches, one of which was to be to a joint session of the state legislature. I was somewhat frustrated because I wanted to spend time with my family, reestablishing ties with my children, and I didn't have time to prepare speeches.

Purely as an expedient measure to extend the duration of my speech to the legislature and to make it more respectable, I decided to quote my poem "Oh Tennessee, My Tennessee" at the end. Lo' and behold, a few weeks later, the legislature decided to adopt the poem as the official poem of the state, since we did not have one.

Memorization activity was a great way to pass the time. I constantly sought to resurrect poems in my mind or get new ones from other POWs. We assigned POWs to be memory banks to retain valuable historical information for dissemination after our release. I served as the memory bank for the names of all known POWs in the system. In our daily communication as we passed information around the camp, we relied almost solely on our memories. Our mental faculties became incredibly sharp.

The average person just does not use his mind as we did, because our whole society is attuned to exchanging information by very rapid means—television, radio, and telephone—most people rarely sit back for many hours and get deeply involved in thought. I think that was one of the most useful things we did as POWs. We learned how much information is in your mind if you really learn how to pull it out.

Our diet as POWs was principally vegetables, supplemented by a small loaf of bread or rice. We were fed twice a day, at about ten o'clock in the morning and four o'clock in the afternoon, purely for their convenience, not ours; but it was amazing how our systems acclimated to that schedule. We were just as hungry at four in the afternoon as we were at ten in the morning. I tried to eat all my food, even though it wasn't very palatable, to maintain my health and strength and I strongly encouraged other POWs to do the same even if they had to force it down. In spite of the limited diet, there were times when we went on fasts to protest certain unsatisfactory conditions in the camp. Fortunately, it did not impair my physical well-being.

Retaining my health and strength enhanced my capability to serve as senior officer in the camp. There were periods where officers in the camp senior to me weren't capable of assuming a leadership role because of being physically debilitated or not willing to get active in the communications system. I really saw the value of the Naval Academy experience, because the principle leaders in North Vietnam were Naval Academy graduates—guys like Stockdale, Denton, McCain, Shumaker. I am not aware of any Naval Academy graduates who did not conduct themselves with great credit as POWs. It really proved to me the value of what we do at the Academy. I believe Naval Academy graduates constituted about 5 percent of the total POW population, and yet received about 50 percent of the very high medals; Stockdale got the Medal of Honor, and several others got the Distinguished Service Medal, like Denton, Shumaker, and myself.

Our captors made a concentrated effort to convert us to their way of thinking. You couldn't really call it brainwashing. I think it was actually motivated by a naive feeling that they could change our views on the war, and they felt this obligation to show us the light and the truth. The antiwar movement in the United States used to send the Vietnamese movies of antiwar demonstrations in the United States. One night in 1970, they herded the POWs from each cell out into the courtyard of our camp and had each group sit far enough away from the other cells so that we couldn't talk among ourselves. It was quite dark, and we could not recognize each other. We were shown a movie of an antiwar demonstration, apparently in the city of San Francisco. I tried to read each of the placards carried by the demonstrators and saw one that said, "Hey, Dick, you can put a man on the moon, but you can't stop the Vietnam War."

Back in our cell I asked Tom Kirk, "Wonder what that means?" Tom said, "Well, Dick must be Richard Nixon." And I said, "Golly you think we really put a man on the moon?" Later, in 1970, a newly shot-down POW confirmed that the United States had in fact put a man on the moon in 1969. That was an example of how the Vietnamese withheld positive information from us. As part of their effort to demoralize us, they only provided adverse information from home to us.

Every cell in our camp had a small, rudimentary speaker with very bad fidelity. For a half hour in the morning and in the afternoon, they would play us their daily Voice of Vietnam broadcast, which must have been an international broadcast. The announcer, Hanoi Hannah as we called her, spoke in very good English with a little bit of an oriental accent. The broadcast was very much oriented to proclaiming their great victories in the war as well as describing all the bad things that were happening in the United States. So for six years I learned about every calamity and adversity that happened in the United States. But most POWs just kind of tuned it out and didn't pay much attention to it. But I always listened to it very carefully, because I wanted to get as much news about the war and the United States as possible even though it was distorted.

When they would talk about the war I found that if you carefully analyzed what they were telling you, you could get a definite feel for how the war was progressing. When they reported U.S. casualties, if you would divide that by four, that was usually pretty correct. Another thing they gave us was a propaganda newspaper, and that, also, told about engagements in the war. I used to read that very carefully, so, surprisingly, from this hour a day of intense anti-American broadcast, coupled with that newspaper, I really knew a lot about how the war had progressed on the ground in South Vietnam.

We got acclimated to the diet and most conditions in the camp, but we were never able to adjust to the heat. Most of us developed severe heat rash, sometimes progressing to heat sores. From about December to March, the heat moderated somewhat and became bearable. We truly suffered in the summer months.

Over the years the guards developed a begrudging respect for the prisoners, because they observed how we hung together and took care of ourselves. We were very self-sufficient. But I think the fact that they saw a lot of POWs take torture to resist doing something made them realize we were courageous individuals. The respect manifested itself in more considerate treatment in the later years.

We knew it would be virtually impossible to escape from Hoa Lo, a maximum security prison surrounded by a moat and high walls covered with broken glass, but we were constantly developing escape plans. When John Dramesi and Ed Atterbury, two air force captains, escaped from another camp in Hanoi called the Zoo—and they'd been preparing for this for years and years—they were captured after traveling several miles down the Red River on the way to the sea. Ed Atterbury apparently died during the torture he received after recapture. John Dramesi, a real tough nut, survived and later described his ordeal. Two other POWs escaped from a small camp but were only out for a brief period. No one ever successfully escaped from North Vietnam to my knowledge.

After the unsuccessful rescue attempt by U.S. special forces troops at the POW camp at Son Tay near the Laos border in November 1970, the Vietnamese moved virtually all U.S. POWs into Hoa Lo prison. From that time until the end of the war, over three hundred POWs were concentrated there, apparently for maximum security. Surprisingly, we learned about the Son Tay rescue attempt fairly soon, because some South Vietnamese army prisoners heard about this and passed this to three Thai prisoners in our prison. From those Thai prisoners, who spoke broken English, we learned about the Son Tay raid—not all the details, but we were able to put two and two together, and assumed this was the reason why all the various POWs were consolidated at Hoa Lo.

Religion was a sustaining element in our life. For the first four years or so, we were not allowed to engage in any group religious activity. Religion was essentially a private endeavor by POWs in individual cells. After the Son Tay raid when the POWs were consolidated in Hao Lo prison, in late 1970, we were placed in large open cells, around forty to fifty POWs in each group. In my particular room we decided to have a group religious service the first Sunday we were together. Robbie Risner and Howie Rutledge were leading the service. The guards observed this through the window of our cell and then came in and removed Risner and Rutledge.

We were very indignant about this happening as it was obviously an over-reaction of the guard to seeing a benign religious service. So, to show our indignation, we decided that those of us who were the senior leadership (grade O-5 and above) would go on a two-day fast. We sent the word around the camp that, in the evening after sunset when things were quiet, on a pre-arranged signal, we would all stand up and, at the top of our lungs, sing in unison *The Star Spangled Banner*. I'm sure our singing was heard throughout Hanoi. The Vietnamese camp officials must have been very embarrassed, and overreacted by taking those of us who were the senior POWs and putting us back to living in the small cells once more. Later on, after they saved face by taking our leaders away, they did start allowing group religious service. So we thought that was one of our great victories in the POW camp.

One night after dark in December 1972, we suddenly heard a tremendous rumbling sound, which persisted for several minutes. Nobody in our group could guess what the sound was. Some speculated that it was an earthquake or an industrial accident. None of us really had any idea that it might be a B-52 bombing raid. The next day in the regular propaganda broadcast, they did reveal that North Vietnam was being attacked by the B-52s. We observed a big change in attitude of the Vietnamese camp personnel from one of confidence and cockiness to a look of concern and anxiety. As soon as dusk came, the next several nights, they stayed in bomb shelters, which they seldom did before. We could clearly see their eagerness to get the war over. I remember some of the camp personnel that spoke English saying, "Why is your country doing to this to us?" You could really see the great impact on their morale by the B-52 bombing.

The POWs had little concern that the B-52 bombs would hit our prison. As military pilots I guess we were eternal optimists, and after enduring the American bombing for many years without any damage to our camp, we felt pretty confident we were safe. The bombing really buoyed our spirits and we became cautiously optimistic that the end of the war was approaching. We had known for several years that negotiations were in progress and we had gotten our hopes up several times, only to have them dashed. So we restrained our emotions.

About a month after the December 1972 bombing, the Vietnamese broadcast to us that on 27 January 1973 a peace treaty had been signed. We were informed that one of the stipulations of the peace treaty was that the POWs would be released and that we would be provided the specific details of the treaty. Later they gave us a mimeographed sheet that listed all the provisions of the treaty, and the details of our release. Though we were overjoyed we still were restrained, guarding against a possible reversal of the situation.

In the final month or so that we were there, they grouped us by the order in which we had been shot down, because one of the stipulations of the peace treaty was that we would be released in that order. As I recall, our release was conducted in four segments. The first was in the early part of February 1973. I was released in the third group on 4 March 1973, and I think there was a final fourth group after that.

Up until the time our release actually occurred, I contained my euphoria; I refused to allow myself to think a lot about the reunion with my family, because, for one thing, I hadn't been allowed to receive any mail from my wife or my children the whole six years I was there. I could hardly conceive what my children looked like. I had received a few letters from my parents that were just six lines—all that were permitted—and they were very noncommittal-type letters, no real substantive information in them. I think over the last three years I was there, I was allowed to send out maybe about ten letters, and they were very short letters to my parents. But I had no idea about the status of my wife and children. I just had the optimistic hope that they were all okay, that the government and my other family members had taken good care of them.

When the release occurred, my group was taken over in buses to the commercial airport in Hanoi, and we all were lined up in the order that we had been shot down. They had taken our striped prison garments and given us new attire to put on, cotton trousers and a simple jacket, and even leather shoes, as I recall. There was a very formal release procedure, where they had a desk with a register; your name was announced and you had to come and check out at the desk, and then an air force officer escorted you out to the C-141 that was waiting. We all got aboard the waiting C-141 transport, and it was not until the airplane got airborne, on the way from Hanoi down to Clark Air Force Base in the Philippines, that we really allowed our exuberance to erupt. We were still somewhat restrained until we saw that coastline pass under us, and we began cheering at the top of our lungs. The ordeal was finally over.

There's a famous picture that appeared in the newspapers of our POW group sitting in the airplane, shouting and yelling as we went "feet wet"— past the coastline. There was an air force public affairs officer on the airplane who told me that his wife was from my hometown of Nashville. I informed him that I remembered the name of her family. He advised me that I would be going to the naval hospital at Memphis. Although I really didn't think seriously about it, I said to myself, "Why am I going down to Memphis, Tennessee, when my family's in the San Diego area?" Because my squadron had been at Miramar, I expected to go back to a hospital in San Diego.

Capt. William P. Lawrence (right) *returns to freedom on 3 March 1973 after six years of confinement and torture in a North Vietnamese prison camp.* (Courtesy Vice Adm. William P. Lawrence, USN [Ret.])

We landed at Clark and received a warm, rousing reception from a large crowd assembled on the ramp. As we came off the airplane, we were greeted by Adm. Noel Gayler, who was commander in chief, Pacific Command, and his deputy, Lt. Gen. Bill Moore, U.S. Air Force. It was getting late in the afternoon and we were taken immediately by bus over to the hospital at Clark. The school children from around the United States, including those at Clark, had prepared welcome home posters that were displayed throughout the hospital.

We remained at Clark only about three days, receiving comprehensive medical tests, uniforms, and personal items. Medically, I appeared okay except for intestinal worms. After six years on a poor diet and living in a dark cell, I needed glasses, which they provided.

At Clark I learned for the first time that my wife, Anne, had divorced me and married an Episcopal minister. I, of course, was devastated. After several hours of grieving, I realized, as I had so many times as a POW, that I simply had to get this behind me and rebuild my life. I had three children who urgently needed my attention and guidance.

A lot of people have asked, "How long did it take you to get acclimated to getting typical American food and sleeping on a soft bed?" I said, "About five seconds." That was a very easy transition. But I had perceived that it was, in some respects, better to be on the type of low-fat diet that we had been on as POWs, so I was determined that I would not resume eating a lot of beef and fatty foods. So, right from the day of release, I began eating principally fruit, vegetables, and salad. That was the thing that I enjoyed the most, that I found I had actually missed the most. From that day on, I basically stayed on a very low-fat diet, very little beef, and I've kept a very low cholesterol, always under two hundred. I think my cholesterol when I came back was about one hundred forty, and now it's around one hundred eighty. So that was one good thing about POW experiences—that it basically gave me some good dietary habits, that I think have been useful to me healthwise.

It was a very brief period at Clark, but we were all eager to be reunited with our families. The navy informed me I would undergo convalescence at the U.S. Naval Hospital, Memphis, because they felt it would be better for me to be close to my parents down in Tennessee. We were flown to the States in air force hospital evacuation aircraft, with a bunk for each POW. We stopped in Hickham in Honolulu about two o'clock in the morning, where we were met by senior military officials; all of my Academy classmates stationed in that area came out, and so we had a nice pleasant visit in the middle of the night, remembering old times.

We proceeded on to Scott Air Force Base in Illinois, and they put six of us into a C-9 for the flight down to Memphis for admission to the hospital. I was the senior one in the group, so I knew it would be expected that the senior POW of the group would make a speech to the assembled people greeting us in Memphis, as had occurred at Clark.

When I arrived there at NAS Memphis, I was escorted by a rear admiral and the commanding officer of the naval air station over to a microphone, in front of all the assembled people. I got to the microphone and directly in front of me were my parents, my two brothers, my three children, and my son's wife. It was a very emotional experience and it was all I could do to keep my composure as I delivered my speech.

We had a wonderful reunion with our families, and then we were taken to the hospital. We met the intelligence team that would debrief us the next several days and we prepared for our convalescence. I think this was one of the few wars where returning POWs were treated as celebrities, and the unfortunate thing about it is that the POWs were, at that time, the only real heroes of that war. Other veterans were treated poorly, and this was a great disappointment to the returning POWs. I think that the country badly wanted to have some heroes from that unpopular war, and it just turned out that the POWs were the ones who were thrust into that position.

Other returning veterans were not even welcomed home. My younger brother had a commission in the army and served in the 82nd Airborne Division in Vietnam. After he finished up his year with the 82nd Airborne, he returned to the Oakland Army Depot, switched immediately into civilian clothes, and flew back to Nashville. My parents met him and he just went back in civilian life as though nothing had happened. I guess most veterans wanted to come back to America quietly, with no fanfare, because there was such an antiwar mood.

Isn't that terrible? A guy just comes back after a hard year in Vietnam, switches to civilian clothes, and if someone saw him on the street, they wouldn't even remember the fact that he'd been in Vietnam. It was a terrible, terrible way that the veterans of that war were treated.

Appendix A
Naval Aviation Organization: 1945–1975

CIVILIAN ADMINISTRATION

Secretary of Defense

President	Secretary	Apptd.
Harry S Truman	James V. Forrestal	1947
	Louis A. Johnson	1949
	George C. Marshall	1950
	Robert A. Lovett	1951
Dwight D. Eisenhower	Charles E. Wilson	1953
	Neil H. McElroy	1957
	Thomas S. Gates, Jr.	1959
John F. Kennedy	Robert S. McNamara	1961
Lyndon B. Johnson	Robert S. McNamara	1963
	Clark M. Clifford	1968
Richard M. Nixon	Melvin R. Laird	1969
	Elliot L. Richardson	1973
	James L. Schlesinger	1973
Gerald R. Ford	James L. Schlesinger	1974
	Donald H. Rumsfeld	1975

Secretary of the Navy

President	Secretary	Apptd.
Harry S Truman	James V. Forrestal	1945
	John L. Sullivan	1947
	Francis P. Matthews	1949
	Dan A. Kimball	1951
Dwight D. Eisenhower	Robert B. Anderson	1953
	Charles S. Thomas	1954
	William B. Franke	1958
John F. Kennedy	John B. Connally, Jr.	1961
	Fred Korth	1961
Lyndon B. Johnson	Paul H. Nitze	1963
	John T. McNaughton	1967
	Paul R. Ignatius	1967
Richard M. Nixon	John H. Chafee	1969
	John W. Warner	1972
Gerald Ford	J. W. Middendorf II	1974

Assistant Secretary of the Navy for Air

Established by Act of Congress 24 June 1926 with title Assistant Secretary of the Navy for Aeronautics. Office vacant 1 June 1932 to 5 September 1941 and 11 September 1941 retitled Assistant Secretary of the Navy for Air. Abolished on 5 February 1959.

Artemus L. Gates	5 Sept. 1941–1 July 1945
John L. Sullivan	1 July 1945–17 June 1946
John Nicholas Brown	12 Nov. 1946–8 Mar. 1949
Dan A. Kimball	19 Mar. 1949–25 May 1949
John F. Floberg	5 Dec. 1949–23 July 1953
James Hopkins Smith	23 July 1953–20 June 1956
Garrison R. Norton	28 June 1956–5 Feb. 1959

NAVAL AVIATION COMMANDS: 1945–1975

Chief of Naval Operations

Fleet Adm. Chester W. Nimitz	15 Dec. 1945–15 Dec. 1947
Adm. Louis E. Denfeld	15 Dec. 1947–2 Nov. 1949
Adm. Forrest P. Sherman	2 Nov. 1949–22 July 1951
Adm. William M. Fechteler	16 Aug. 1951–17 Aug. 1953
Adm. Robert B. Carney	17 Aug. 1953–16 Aug. 1955
Adm. Arleigh A. Burke	17 Aug. 1955–1 Aug. 1961
Adm. George W. Anderson	1 Aug. 1961–31 July 1963
Adm. David L. McDonald	1 Aug. 1963–31 July 1967
Adm. Thomas H. Moorer	1 Aug. 1967–1 July 1970
Adm. Elmo R. Zumwalt, Jr.	1 July 1970–30 June 1974
Adm. James L. Holloway	1 July 1974–1 July 1978

Deputy Chief of Naval Operations (Air Warfare)

Established by the Secretary of the Navy, 18 August 1943, as Deputy Chief of Naval Operations (Air). Changed to Air Warfare, 15 July 1971.

Vice Adm. Aubrey W. Fitch	1 Aug. 1944–14 Aug. 1945
Vice Adm. Marc A. Mitscher	14 Aug. 1945–15 Jan. 1946
Vice Adm. Arthur W. Radford	15 Jan. 1946–22 Feb. 1947
Vice Adm. Donald B. Duncan	6 Mar. 1947–20 Jan. 1948
Vice Adm. John D. Price	20 Jan. 1948–6 May 1949
Vice Adm. Calvin T. Durgin	16 May 1949–25 Jan. 1950
Vice Adm. John H. Cassady	25 Jan. 1950–31 May 1952
Vice Adm. Matthias B. Gardner	31 May 1952–16 Mar. 1953
Vice Adm. Ralph A. Ofstie	16 Mar. 1953–3 Mar. 1955
Vice Adm. Thomas S. Combs	11 Apr. 1955–1 Aug. 1956
Vice Adm. W. V. Davis, Jr.	1 Aug. 1956–22 May 1958
Vice Adm. Robert B. Pirie	26 May 1958–1 Nov. 1962
Vice Adm. William A. Schoech	14 Nov. 1962–1 July 1963
Vice Adm. John S. Thach	8 July 1963–25 Feb. 1965
Vice Adm. Paul H. Ramsey	31 Mar. 1965–1 Oct. 1966
Vice Adm. Thomas F. Connolly	1 Nov. 1966–31 Aug. 1971
Vice Adm. M. F. Weisner	1 Sept. 1971–4 Aug. 1972
Vice Adm. W. D. Houser	5 Aug. 1972–30 Apr. 1976

APPENDIX A

Chief of the Bureau of Aeronautics

Established by Act of Congress, 12 July 1921, and merged 1 December 1959 with the bureau of Ordnance to form the Bureau of Naval Weapons.

Rear Adm. Dewitt C. Ramsey	7 Aug. 1943–1 June 1945
Rear Adm. Harold B. Sallada	1 June 1945–1 May 1947
Rear Adm. Alfred M. Pride	1 May 1947–1 May 1951
Rear Adm. Thomas S. Combs	1 May 1951–30 June 1953
Rear Adm. Apollo Soucek	30 June 1953–4 Mar. 1955
Rear Adm. James S. Russell	14 Mar. 1955–15 July 1957
Rear Adm. Robert E. Dixon	15 July 1957–1 Dec. 1959

Chief of the Bureau of Naval Weapons

Established by Act of Congress, 18 August 1959, merging the bureaus of Ordnance and Aeronautics. Abolished by reorganization of 1 May 1966, which assigned elements to three new commands: Naval Air Systems Command, Naval Ordnance Systems Command, and Naval Electronic Systems Command.

Rear Adm. Paul D. Stroop	10 Sept. 1959–29 Oct. 1962
Rear Adm. Kleber S. Masterson	27 Nov. 1962–24 Mar. 1964
Rear Adm. Allen M. Shinn	28 May 1964–1 May 1966

Commander, Naval Air Systems Command

Established by a reorganization of the Navy Department effective 1 May 1966.

Rear Adm. Allen M. Shinn	1 May 1966–1 Sept. 1966
Rear Adm. Robert L. Townsend	1 Sept. 1966–20 Feb. 1969
Rear Adm. T. J. Walker III	20 Feb. 1969–1 Apr. 1971
Rear Adm. T. R. McClellan	1 Apr. 1971–31 Aug. 1973
Vice Adm. Kent L. Lee	31 Aug. 1973–29 Aug. 1976

Director of Marine Corps Aviation

Maj. Gen. Field Harris	18 July 1944–24 Feb. 1948
Maj. Gen. William J. Wallace	24 Feb. 1948–1 Sept. 1950
Brig. Gen. Clayton C. Jerome	1 Sept. 1950–1 Apr. 1952
Lt. Gen. William O. Brice	1 Apr. 1952–31 July 1955

Lt. Gen. Christian F. Schilt	1 Aug. 1955–31 Mar. 1957
Lt. Gen. Verne J. McCaul	1 Apr. 1957–2 Dec. 1957
Maj. Gen. Samuel S. Jack	14 Jan. 1958–20 Feb. 1958
Maj. Gen. John C. Munn	21 Feb. 1958-14 Dec. 1959
Maj. Gen. Arthur F. Binney	15 Dec. 1959–10 Sept. 1961
Col. Keith B. McCutcheon	11 Sept. 1961–17 Feb. 1962
Col. Marion E. Carl	18 Feb. 1962–4 July 1962
Brig. Gen. N. J. Anderson	5 July 1962–20 Oct. 1963
Maj. Gen. L. B. Robertshaw	21 Oct. 1963–15 June 1966
Maj. Gen. K. B. McCutcheon	15 June 1966–18 Feb. 1970
Maj. Gen. Homer S. Hill	19 Feb. 1970–24 Aug. 1972
Maj. Gen. Edward S. Fris	25 Aug. 1972–27 Aug. 1974
Brig. Gen. Philip D. Shutler	28 Aug. 1974–January 1975
Maj. Gen. V. A. Armstrong	January 1975–21 Aug. 1975
Lt. Gen. T. H. Miller, Jr.	22 Aug. 1975–29 June 1979

Commander, Naval Air Force, Atlantic Fleet

Established 1 January 1943 as an administrative command replacing the commands Carriers, Atlantic Fleet and Fleet Air Wing, Atlantic Fleet. The original title, Air Force, Atlantic Fleet, was changed 30 July 1957 to Naval Air Force, Atlantic Fleet.

Vice Adm. P. N. L. Bellinger	20 Mar. 1943–2 Feb. 1946
Vice Adm. Gerald F. Bogan	2 Feb. 1946–December 1948
Vice Adm. Felix B. Stump	December 1948–11 May 1951
Vice Adm. John J. Ballentine	11 May 1951–1 May 1954
Vice Adm. F. G. McMahon	1 May 1954–29 May 1956
Vice Adm. William L. Rees	29 May 1956–30 Sept. 1960
Vice Adm. F. O'Beirne	30 Sept. 1960–30 Sept. 1963
Vice Adm. Paul H. Ramsey	30 Sept. 1963–31 Mar. 1965
Vice Adm. Charles T. Booth	31 Mar. 1965–28 Feb. 1969
Vice Adm. Robert L. Townsend	1 Mar. 1969–29 Feb. 1972
Vice Adm. F. H. Michaelis	29 Feb. 1972–14 Feb. 1975
Vice Adm. Howard E. Greer	14 Feb. 1975–31 Mar. 1978

Commander, Naval Air Force, Pacific Fleet

Established 1 September 1942 as an administrative command replacing the commands Carriers, Pacific Fleet and Patrol Wings, Pacific Fleet. The title U.S. Naval Air Forces, Pacific Fleet was

APPENDIX A

changed 14 October 1942 to Air Force, Pacific Fleet and 30 July 1957 to Naval Air Force, Pacific Fleet.

Rear Adm. George D. Murray	17 Aug. 1944–20 July 1945
Rear Adm. A. E. Montgomery	20 July 1945–31 Aug. 1946
Vice Adm. John D. Price	31 Aug. 1946–5 Jan. 1948
Vice Adm. Harold B. Sallada	5 Jan. 1948–1 Oct. 1949
Vice Adm. Thomas L. Sprague	1 Oct. 1949–1 Apr. 1952
Vice Adm. Harold M. Martin	1 Apr. 1952–1 Feb. 1956
Vice Adm. Alfred M. Pride	1 Feb. 1956–30 Sept. 1959
Vice Adm. C. E. Ekstrom	21 Oct. 1959–30 Nov. 1962
Vice Adm. Paul D. Stroop	30 Nov. 1962–30 Oct. 1965
Vice Adm. Thomas F. Connolly	30 Oct. 1965–1 Nov. 1966
Vice Adm. Allen M. Shinn	1 Nov. 1966–31 Mar. 1970
Vice Adm. William F. Bringle	31 Mar. 1970–28 May 1971
Vice Adm. T. J. Walker III	28 May 1971–31 May 1973
Vice Adm. Robert B. Baldwin	31 May 1973–12 July 1976

Appendix B
Glossary of Terms and Abbreviations

A/AAA: antiaircraft fire; antiaircraft artillery

AAW: antiair warfare

Afterburner: provides a surge of power by feeding raw fuel into a jet's hot exhaust; used to gain maximum power for increasing speed or energy levels; referred to as burner or AB

Air officer: the ship's officer responsible for aviation matters in an aircraft carrier

Air plot: the air operations control center aboard a carrier

Alpha strike: a large offensive air strike, involving all the carrier air wing's assets—fighters, attack, refueling, etc.

Angels: altitude of an aircraft in thousands of feet (e.g., "Angels three five" is 35,000 feet)

Angle of attack: the angle between the chord line of an airfoil and the relative airflow, normally the immediate flight path of the airplane

Angled deck: the flight deck of an aircraft carrier extended diagonally from the port side of the ship, so that airplanes can take off and land without interference to or from airplanes parked on the bow

APQ-72: the missile control system in the F-4 Phantom II fighter

Arresting gear: the arrangement of wires on a carrier flight deck that stops an airplane after the airplane's tailhook has engaged it

ASW: antisubmarine warfare

AWACS: Airborne Warning and Control System

Ball (also "meatball"): the optical landing aid that guides pilots while landing aboard ship

Bandit: an enemy aircraft

Barricade: a collapsible arrangement of vertical webbing rigged on an aircraft carrier in an emergency to arrest airplanes unable to make a normal arrested landing

Barriers: collapsible fences on a carrier flight deck that stop those airplanes whose hooks have missed the arresting gear; before the advent of the angled deck

Blackshoes: surface or ship personnel; brownshoes refer to aviators

Blip: a target indication on a radar scope

BN: A-6 or A-3 bombardier-navigator

Bogey: an unidentified aircraft

Bolter: a takeoff following unsuccessful (missed) arrestment aboard a carrier with an angled deck

Break: a rapid hard turn to avoid SAM or aircraft, or a prescribed pattern when an aircraft returns to the ship (heading into the break)

BuAer: Bureau of Aeronautics

Bullpup: an air-to-surface missile fired and directed by radio control to the target; normally fired pointing down

CAG: air group commander; when air group designation was changed to air wing, the CAG title remained part of the vocabulary

CAP: combat air patrol

Carrier task force: the force of aircraft carriers and supporting ships

Cat: catapult; a steam-actuated mechanism used to hurl (launch) aircraft from the deck of an aircraft carrier

CBU: cluster bomb unit

CEF: Cuban Expeditionary Force during the Bay of Pigs crisis

Chopper: slang for helicopter

CIA: Central Intelligence Agency

CIC: Combat Information Center; the space in a ship containing radar, plotting and communication gear

Chief of staff: the captain or admiral who assists an admiral, as his second in command, especially in supervising his staff

CinCLant: Commander in Chief, Atlantic

CinCPac: Commander in Chief, Pacific

CinCPacFlt: Commander in Chief, Pacific Fleet; in charge of all naval forces in the Pacific Fleet; reports directly to CinCPac, a navy four-star admiral responsible for all U.S. forces in the Pacific

GLOSSARY OF TERMS AND ABBREVIATIONS

CinCSouth: United States Commander in Chief, Southern Command

CNO: Chief of Naval Operations; the top uniformed officer in the navy

CO: commanding officer

COD: carrier on-board delivery; usually C-l/C-2 aircraft that deliver mail, supplies, and personnel to carriers while ships are under way

ComCarDiv: Commander, Carrier Division

ComDesRon: Commander, Destroyer Squadron

ComNavAirLant: Commander, Naval Air Forces, Atlantic Fleet; frequently abbreviated to AirLant

ComNavAirPac: Commander, Naval Air Forces, Pacific Fleet; frequently abbreviated to AirPac

Conn: control of a ship's movements; to guide or pilot a ship

Cruise: a tour of sea duty

CTF: Commander, Task Force; CTF-77 directed Yankee Station activity, with several carrier groups under the command

Cut: a mandatory signal from LSO for the pilot to land on the carrier

CV: aircraft carrier

CVA: attack aircraft carrier

CVAN/CVN: attack aircraft carrier, nuclear-powered

CVE: escort carrier

CVG/CVW: carrier air group/air wing; aircraft of a carrier, made up of squadrons; CVG no longer used

CVL: light carrier

CVS: antisubmarine aircraft carrier

DCNO (Air): Deputy Chief of Naval Operations for Air

Deep draft: a large, oceangoing vessel; naval aviators usually command a deep draft ship before assuming command of carrier

Det: detachment; not a full squadron, but a portion thereof

DEW: Distant Early Warning; defensive line of radar stations at about the 70th parallel on the North American continent

DFC: Distinguished Flying Cross

Division: the basic administrative unit into which men are divided on board ship, in aircraft squadrons, or at shore activities; also, a tactical subdivision of a squadron of ships or aircraft (four)

Dixie Station: the point off South Vietnam in the Gulf of Tonkin designated for aircraft carrier operations in South Vietnam, Laos, and Cambodia

DLG: guided missile frigate

DOD: Department of Defense

ECM: electronic countermeasures

Essex-*class carrier:* a class of World War II carrier, displacing 27,500 tons, which was modernized after World War II to accommodate heavier, high-performance aircraft

FAA: Federal Aviation Administration

FAC: forward air controller; pinpoints enemy concentrations to ground personnel or incoming air strikes; FAC usually flew in a light plane

Feet dry/feet wet: a pilot has reached land (dry), or water (wet)

Flag: admiral rank

Flag bridge: bridge of a ship used by a flag officer and his staff

Flag lieutenant: personal aide to a flag officer

Flag plot: the enclosed tactical and navigational center used by a flag officer and his staff in exercising tactical command of ships and aircraft

Flag secretary: the personal aide to a flag officer who handles the paperwork of a staff

Flak: antiaircraft fire

Flat spin: a flight condition of rotation about the vertical axis while the longitudinal axis is inclined downward less than 45 degrees with the horizontal or level plane

GCA: ground-controlled approach; a form of directed approach used to control the flight path of an aircraft by radar during a landing approach to an airport in low-visibility conditions

GCI: ground-controlled intercept; system whereby aircraft are directed to an interception course by a radar station on the ground

General quarters: the condition of maximum readiness for combat with the crew at battle stations

G-suit: the anti-G suit worn by pilots; a system of air bladders that compress legs and abdomen so the blood does not pool in the lower extremities during high-performance maneuvers; prevents loss of blood to the brain

Hanoi Hilton: the nickname for the Hoa Lo prison, which housed the majority of American prisoners of war; located in the middle of Hanoi

Harpoon: air-launched anti-ship missile

Heartbreak Hotel: located at Hoa Lo prison; a receiving station for new POWs

HU: navy helicopter squadron, as in HU-2

IATU: Instructors Advanced Training Unit
ICBM: intercontinental ballistic missile
IFF: Identification, Friend or Foe; an electronic identification system used
    by aircraft to identify themselves
In-country: within South Vietnam
IOIC: Integrated Operational Intelligence Center
IR: infrared; applicable to the heat-seeking Sidewinder missile
Iron Hand: missions against SAM sites
Iron Triangle: the highly defended area between Haiphong, Hanoi, and
    Thanh Hoa

JAG: Judge Advocate General; the senior legal specialist in the navy
JCS: Joint Chiefs of Staff
Jink: to constantly change altitude and direction to present an irregular flight
    pattern, making it difficult for gunners to track the plane
JO: junior officer, lieutenant and below
JOC: Joint Operations Center

LPH: amphibious assault ship
LSO: landing signal officer; the officer stationed on a platform at the aft end
    of a carrier who assists pilots in carrier landings
LTV: Ling-Temco-Vought

Mach number: based on the ratio of the speed of an aircraft to the speed at
    which sound can travel under the same conditions; a plane traveling at
    Mach one is traveling at the local speed of sound
MACV: Military Assistance Command, Vietnam
Midway-*class carrier: Midway, Franklin D. Roosevelt,* and *Coral Sea*;
    heavily armed and armored carriers, modernized after World War II
    for jet operations
MiG: Mikoyan-Gurevich (Soviet aircraft)
MiG CAP: standing patrol over the fleet or strike force to protect against any
    threat by enemy aircraft in Vietnam
Mirror landing system: the optical guidance system used to direct carrier
    aircraft to a successful landing; replaced by Fresnel lens

NAS: Naval Air Station
NATO: North Atlantic Treaty Organization
Naval Air Systems Command: the command responsible for the material
    support of naval aviation

NTDS: Naval Tactical Data System

OJCS: Organization of the Joint Chiefs of Staff

OOD/SDO: Officer of the Deck/Squadron Duty Officer; the officer who is responsible for handling the daily routine of the ship/squadron

OpNav: Office of the Chief of Naval Operations; OP-05 (DCNO for Air) is a division within OpNav

ORI: operational readiness inspection

OSD: Office of the Secretary of Defense

Over the beach: over land, feet dry

PACAF: Pacific Air Force

PERT: Program Evaluation and Review Technique; highly acclaimed system used to manage the Polaris Fleet Ballistic Missile Submarine program

Pipper: the center dot in gun sight

PIRAZ: Positive Identiflcation Radar Advisory Zone; a ship stationed on a rotational basis in the Tonkin Gulf, providing navigational, weather, and threat information for U.S. aircraft

Pop-up point: the geographic point at which a pilot starts his climb to gain altitude for dive (bomb delivery)

POW: prisoner of war

PRC: People's Republic of China

Probe-and-drogue: the system of aerial refueling whereby the receiver airplane inserts a refueling probe into the tanker's drogue, or basket, at the end of a hose through which fuel is pumped

Ready Room: the compartment on a carrier where aircrew gather for flight briefing, training, etc.

Recce: shortened form for reconnaissance

Regulus I: a simple, subsonic, medium-range, surface-to-surface missile for operation from all types of naval vessels

Regulus II: a supersonic, surface-to-surface missile for operation from submarines

RESCAP: rescue combat air patrol

RIO (or RO): radar intercept officer; backseater in the F-4 Phantom

RN: Royal Navy

ROE: rules of engagement

Roger: the LSO signal with arms (paddles) extended horizontally to either side, meaning the landing approach is satisfactory

ROK: Republic of Korea

Route packages: geographic areas established in North Vietnam for purposes of greater control of strikes by U.S. air forces

RVAH: navy heavy reconnaissance squadron (RA-5C Vigilante)

SAC: Strategic Air Command

SACLant: Supreme Allied Commander, Atlantic

SAM: surface-to-air missile; a generic term, but usually referring to the Soviet-built SA-2 Guideline

SAR: search and rescue

SARCAP: search and rescue combat air patrol

Section: with aircraft, generally a unit of two planes

Shrike: anti-radar missile fired from aircraft at SAM sites

Shuttle: the device to which a carrier aircraft is attached to accelerate the plane down the catapult track

Sidewinder: a heat-seeking, air-to-air missile

Sixth Fleet: the U.S. fleet in the Mediterranean

Sled: the net trailed behind a ship in which a hook on a seaplane's pontoon engages as the seaplane is recovered with the ship under way

Spad: the nickname for Douglas A-1 Skyraider

Sparrow III: the all-weather, semiactive, radar homing air-to-air missile

Splash: enemy aircraft destroyed

Squadron: with aircraft, a unit of 12 to 24 planes

Strike: a combat flight against ground or ship targets

Styx: the Soviet surface-to-surface cruise missile, normally carried on fast patrol boats

TACAN: tactical air navigation system

Tally-ho: sight contact on a target by a combat aircraft

Talos: the shipborne, long-range, surface-to-air guided missile

Tandem: one behind the other

TARCAP: target combat air patrol; fighters tasked with providing escort protection for the strike force

Tartar: the shipborne, medium-range, surface-to-air guided missile

TBS: talk between ships (radio)

Terrier: the shipborne, short-range, surface-to-air guided missile

TF: task force

TG: task group

Third Fleet: the U.S. fleet in the eastern Pacific

Tin can: slang for destroyer

Touch-and-go landing: a landing during which the pilot applies full takeoff power after touching down, with the intent of taking off, rather than coming to a full stop

TPS: Test Pilot School

Trap: an arrested landing aboard a carrier

UHF: ultrahigh frequency

Unrep: underway replenishment; process of supplying food, fuel, stores, ammunition, and personnel to fleet combatant ships while under way

VA: navy attack squadron (A-I Skyraider, A-4 Skyhawk, A-6 Intruder, A-7 Corsair)

VAH: navy heavy attack squadron (A-3B Skywarrior)

VAP: navy heavy photographic squadron (RA-3B Skywarrior)

VAQ: navy tactical electronic warfare squadron (EKA-3 Skywarrior, later EA-6B Prowler)

VC: navy composite squadron; originally special mission aircraft such as night attack (AD-4N) or night fighter (F4U-5N) aircraft

VCNO: Vice Chief of Naval Operations

Vector: an aircraft's heading; to direct, to give a heading

VF: navy fighter squadron (F-4 Phantom or F-8 Crusader)

VFP: navy light photographic squadron (RF-8 Crusader)

VHF: very high frequency

VMA: marine attack squadron (A-4 Skyhawk)

VMA(AW): marine all-weather attack squadron (A-6 Intruder)

VMCJ: marine composite reconnaissance squadron (EF-10B Skyknight, RF-4B Phantom, EA-6A Intruder, and EA-6B Prowler, providing photographic and ECM services)

VMFA: marine fighter and attack squadron (F-4 Phantom)

VMF(AW): marine all-weather fighter squadron (F-8 Crusader, later F-4 Phantom)

VMO: marine observation squadron (UH-1 helicopter, later OV-10 Bronco)

VP: navy patrol squadron (P-5 Marlin, P-2 Neptune, P-3 Orion)

VS: navy antisubmarine squadron (S-2 Tracker, S-3 Viking)

Walleye: the TV-guided air-to-surface glide bomb; the pilot could see his target through the missile's TV eye; introduced in March 1967

Waveoff: a mandatory signal from the LSO to the pilot not to land

XO: executive officer

Yankee Station: the position in the Gulf of Tonkin south of Hainan Island for carrier strikes into North Vietnam

Zoo: the nickname for a POW compound in the North Vietnam prison system on the southwestern edge of Hanoi

Zuni: a 5-inch, air-to-ground unguided rocket

AIRCRAFT MENTIONED IN THE TEXT

U.S. Navy

Note: A unified designation system was placed in effect in September 1962 for U.S. military aircraft.

Antisubmarine

S2F	Grumman	Tracker	To S-2, 1962
S-3	Lockheed	Viking	—

Attack

SBC	Curtiss	Helldiver	Pre-World War II
AD	Douglas	Skyraider	To A-1, 1962
A3D	Douglas	Skywarrior	To A-3, 1962
A4D	Douglas	Skyhawk	To A-4, 1962
A2F	Grumman	Intruder	To A-6, 1962
A3J	North American	Vigilante	To A-5, 1962
A-7	LTV	Corsair II	F-8 derivative

Cargo/Transport

C-1	Grumman	Trader	E-1 derivative
C-2	Grumman	Greyhound	E-2 derivative

Fighter

F3D	Douglas	Skyknight	To F-10B, 1962
F4D	Douglas	Skyray	To F-6A, 1962
F5D	Douglas	Skylancer	Experimental
F-111B	General Dynamics	—	Navy version of TFX
F6F	Grumman	Hellcat	World War II
F7F	Grumman	Tigercat	Pre-1962
F8F	Grumman	Bearcat	Pre-1962
F9F	Grumman	Panther/Cougar	To F-9, 1962

F11F-1F	Grumman	Tiger	Modified F11F w/J79 engine.
F-14	Grumman	Tomcat	Variable geometry
FH	McDonnell	Phantom	FD redesignated
F2H	McDonnell	Banshee	To F-2, 1962
F3H	McDonnell	Demon	To F-3, 1962
F4H	McDonnell	Phantom II	To F-4, 1962
F-18	McDonnell Douglas	Hornet	Developed from Northrop YF-17
FJ	North American	Fury	To F-1, 1962
FR	Ryan	Fireball	Prop/jet
F4U	Vought	Corsair	World War II
F7U	Vought	Cutlass	Tailless
F8U	Vought	Crusader	To F-8, 1962

Helicopter

HTL	Bell	—	To H-13, 1962
HRP	Piasecki	Flying Banana	Model PV-3
H-53	Sikorsky	Sea Stallion	S-65; USMC
HO3S	Sikorsky	—	To H-19, 1962

Observation/Scout

SC	Curtiss	Seahawk	Last navy scout
OS2U	Vought	Kingfisher	Pre-1962

Patrol

P2V	Lockheed	Neptune	To P-2, 1962
P3V	Lockheed	Orion	To P-3, 1962
P6M	Martin	Seamaster	Jet-powered

Trainer

TO/TV	Lockheed	Shooting Star	To T-33, 1962

Transport

SNB	Beech	Kansan	To UC-45, 1962

U.S. Air Force

B-52	Boeing	Stratofortress	Strategic bomber
B-36	Convair	—	Strategic bomber
B-26	Douglas	Invader	Medium bomber

GLOSSARY OF TERMS AND ABBREVIATIONS

C-141	Lockheed	Starlifter	Strategic transport
EC-121	Lockheed	Super Constellation	Special electronics installation
F-104	Lockheed	Starfighter	Mach II fighter
U-2	Lockheed	—	High altitude recce
C-9	McDonnell Douglas	Nightingale	Aeromedical transport
F-15	McDonnell Douglas	Eagle	Fighter
F-86	North American	Sabrejet	Fighter

Miscellaneous

MiG	Mikoyan-Gurevich	—	Soviet fighter series
X-15	NorthAmerican	—	High performance research
SR/45	Saunders Roe	Princess	British ten-engined flying boat

Appendix C
Selected Bibliography

ORAL HISTORIES
Note: Unless otherwise indicated, all oral histories were conducted by John T. Mason, Jr., and are a part of the U.S. Naval Institute Oral History Collection.

Anderson, Adm. George W. 1981.

Bogan, Vice Adm. Gerald F. 1969. Interviewed by Comdr. Etta-Belle Kitchen.

Burke, Adm. Arleigh A. 1979.

Connolly, Vice Adm. Thomas F. 1977.

Dennison, Adm. Robert Lee. 1973.

Foley, Rear Adm. Francis D. 1985. Interviewed by Paul Stillwell.

Griffin, Adm. Charles D. 1972.

Hawkins, Capt. Arthur R. 1983. Interviewed by Paul Stillwell.

Hyland, Adm. John J. 1984.

Johnson, Adm. Roy S. 1980.

Lawrence, Vice Adm. William P. 1990. Interviewed by Paul Stillwell.

Lee, Vice Adm. Kent. 1987. Interviewed by Paul Stillwell.

Miller, Vice Adm. Gerald E. 1976.

Pirie, Vice Adm. Robert B. 1974.

Russell, Adm. James S. 1974.

Thach, Adm. John S. 1971. Interviewed by Comdr. Etta-Belle Kitchen.

BOOKS

Abel, Elie. *The Cuban Missile Crisis.* New York: Lippincott, 1966.

Art, Robert J. *The TFX Decision: McNamara and the Military.* Boston: Little, Brown, 1968.

Cagle, Comdr. Malcolm W., USN, and Comdr. Frank A. Manson, USN. *The Sea War in Korea.* Annapolis: Naval Institute Press, 1957.

Chayes, Abram. *The Cuban Missile Crisis.* New York: Oxford University Press, 1974.

Coulam, Robert F. *Illusions of Choice.* Princeton: Princeton University Press, 1977.

Enthoven, Alain C., and K. Wayne Smith. *How Much Is Enough? Shaping the Defense Program, 1961–1969.* New York: Harper and Row, 1971.

Field, James A., Jr. *History of United States Naval Operations Korea.* Washington, D.C.: Government Printing Office, 1962.

Green, William. *War Planes of the Second World War: Fighters, Vol. 4.* Garden City, NY: Doubleday, 1961.

———. *The World Guide to Combat Planes.* 2 vols. Garden City, NY: Doubleday, 1967.

Halperin, Morton H. *Defense Strategies for the Seventies.* Boston: Little, Brown, 1971.

Karnow, Stanley. *Vietnam: A History.* New York: Viking Press, 1983.

Kennedy, Robert F. *Thirteen Days.* New York: Norton, 1969.

Larkins, William T. *U.S. Marine Corps Aircraft 1914–1959.* Concord, CA: Aviation History Publications, 1961.

Marolda, Edward J. *Illustrated History of Carrier Operations—The Vietnam War.* New York: Bantam Books, 1987.

——— and Oscar P. Fitzgerald. *The U.S. Navy and Vietnam Conflict, Volume II, From Military Assistance to Combat, 1959 to 1965.* Washington, DC: Government Printing Office, 1986.

Matt, Paul R. *United States Navy and Marine Corps Fighters 1918–1962.* Los Angeles: Aero Publishers, 1962.

McNamara, Robert S. *The Essence of Security: Reflections in Office.* New York: Harper and Row, 1968.

Mersky, Peter B., and Norman Polmar. *The Naval Air War in Vietnam.* Baltimore: Nautical and Aviation Publishing Company of America, 1986.

Millis, Walter, with E. S. Duffield. *The Forrestal Diaries.* New York: Viking Press, 1951.

APPENDIX C

Montross, Lynn, et al. *U.S. Marine Operations in Korea 1950–1953*. 4 vols. Washington, DC: Government Printing Office, 1954–62.

Naval Historical Center. *A Short History of the United States Navy and the Southeast Asian Conflict, 1950–1975*. Department of the Navy, 1984.

Polmar, Norman. *Aircraft Carriers*. New York: Doubleday, 1969.

Potter, E. B. *Nimitz*. Annapolis: Naval Institute Press, 1976.

——— and Chester Nimitz, eds. *Sea Power: A Naval History*. Englewood Cliffs, NJ: Prentice-Hall, 1960.

Raymond, Jack. *Power at the Pentagon*. New York: Harper and Row, 1964.

Reynolds, Clark G. *The Fast Carriers: The Forging of an Air Navy*. Huntington, NY: Krieger, 1978.

———. *Admiral John H. Towers: The Struggle for Naval Air Supremacy*. Annapolis: Naval Institute Press, 1991.

Sharp, Adm. U. S. G., USN (Ret.). *Strategy For Defeat, Vietnam in Retrospect*. San Rafael, CA: Presidio Press, 1978.

Simmons, Brig. Gen. E. H., USMC (Ret.). *The United States Marines 1775–1975*. New York: Viking Press, 1974.

Sorensen, Theodore C. *Kennedy*. New York: Harper and Row, 1965.

Stockdale, James, and Sybil Stockdale. *In Love and War*. New York: Harper and Row, 1984.

Swanborough, Gordon, and Peter M. Bowers. *United States Naval Aircraft Since 1911*. New York, 1968; Annapolis: Naval Institute Press, 1976.

Taylor, Maxwell. *The Uncertain Trumpet*. New York: Harper and Row, 1959.

Taylor, Theodore. *The Magnificent Mitscher*. New York: Norton, 1954.

Tillman, Barrett. *Corsair: The F4U in WW II and Korea*. Annapolis: Naval Institute Press, 1979.

———. *MiG Master: The Story of the F-8 Crusader*. Annapolis: Nautical and Aviation Publishing Company of America, 1980.

U.S. Navy, DCNO (Air). *United States Naval Aviation, 1910–1970*. Washington, DC: Government Printing Office, 1970.

———, Naval History Division. *Dictionary of American Naval Fighting Ships, Vol. 2*. Washington, D.C.: Government Printing Office, 1963.

Van Wyen, Adrian O., and Lee M. Pearson. *United States Naval Aviation 1910–1960*. Washington, DC: Government Printing Office, 1961.

Yarmolinsky, Adam. *The Military Establishment: Its Impact on American Society*. New York: Harper and Row, 1971.

Index

INDEX

INDEX

About the Author

A graduate of the U.S. Naval Academy's class of 1950, Capt. E. T. "Tim" Wooldridge was designated a naval aviator in April 1952. As a test pilot at the Naval Air Test Center, Patuxent River, Maryland, he flew every type of aircraft in the U.S. Navy in the late 1950s. He later became commanding officer of a carrier-based fighter squadron and executive officer of the aircraft carrier USS *Forrestal*. Captain Wooldridge also attended the National War College and served several tours of duty on the strategic plans and policy staffs of the Navy Department and Joint Chiefs of Staff.

In 1990 he was appointed Ramsey Fellow and Aviation Historian for the National Air and Space Museum and served in that post until December 1994. During that time he wrote a number of books for the Smithsonian Institution Press, including *Images of Flight* (1983), *Focus on Flight* (1985), and a companion volume to this one, titled *Carrier Warfare in the Pacific: An Oral History Collection,* for the press's "History of Aviation" series.

Captain Wooldridge has written for the Naval Institute's *Naval History* magazine, and he is currently coauthoring the official autobiography of Adm. Thomas H. Moorer, to be published by the Naval Institute Press in spring 1996.